FEB 2 6 2021

Brain Inflamed

ALSO BY KENNETH BOCK, M.D.

Healing the New Childhood Epidemics

The Germ Survival Guide

The Road to Immunity

Natural Relief for Your Child's Asthma

Brain Inflamed

Uncovering the Hidden Causes of Anxiety,
Depression, and Other Mood Disorders in
Adolescents and Teens

Kenneth Bock, M.D.,
FAAFP, FACN, CNS

HARPER WAVE

An Imprint of HarperCollins*Publishers*

The information in this book has been carefully researched by the author and is intended to be a source of information only. While the methods contained herein can and do work, readers are urged to consult with their physicians or other professional advisors to address specific medical or other issues that may be causing an inability to "stick with it." The author and the publisher assume no responsibility for any injuries suffered or damages or losses incurred during or as a result of the use or application of the information contained herein.

Names and identifying characteristics of some individuals have been changed to preserve their privacy and some individuals are composites.

HarperCollins books may be purchased for educational, business, or sales promotional use. For information, please email the Special Markets Department at SPsales@harpercollins.com.

FIRST EDITION

Designed by Nancy Singer

Library of Congress Cataloging-in-Publication Data has been applied for.

ISBN 978-0-06-297087-9

21 22 23 24 25   LSC   10 9 8 7 6 5 4 3 2 1

This book is dedicated to the memory of my best friend, my "buddy" David Marell (1945–2019), the quintessential Teen Whisperer. I miss your friendship, wisdom, and counsel, and the thirty-five years of laughter and love we shared, every day.

Contents

Contents

Foreword

I've known Kenneth Bock for decades, and in that time, I've continually watched him diagnose medical mysteries that have stumped many other practitioners. His interventions have altered the trajectory of his patients' lives and have truly improved outcomes for so many kids suffering from psychological disorders. His skill lies in his ability to accurately home in on—and treat—the root causes of the problem. As you will soon learn, many of the conditions that present in adolescents and teens as psychological in nature are, in fact, physiological. Such conditions are, at least in part, a result of chronic, systemic inflammation that has traveled to the brain.

Inflammation is a fundamental biological process that is essential for our survival. It is inflammation and its myriad expressions that allow our bodies to respond to confrontations from potentially harmful interactions with pathogenic organisms, toxins, traumatic events, and other circumstances that have the potential to prove threatening.

But over the last several decades it has become clear that this life-supportive inflammatory response may persist unchecked and actually pave the way for damaging events across a wide spectrum of bodily tissues. We now recognize that so many of our most pervasive, chronic, degenerative conditions are united in terms of their causation by persistent, unrestrained inflammation. Inflammation is now recognized as playing a central role in coronary artery disease, type 2 diabetes, Alzheimer's disease, depression, and even cancer. So, clearly, we need to explore the role of inflammation in the human body as being far more nuanced than simply friend or foe.

Whether supportive or destructive, the inflammatory response is certainly complex. And while the common initiators described above are generally well understood, what is less universally appreciated is the role of lifestyle choices in regulating inflammation and its downstream effects. Our white blood cells and the various chemical mediators they produce are deeply involved in the inflammatory process, and they can of course be activated by a bee sting or an infection in the skin. But fundamental to the new understanding of the role of persistent inflammation in so many chronic debilitating conditions is the recognition that these very same pathways may be triggered into action by our daily decisions, including inappropriate food choices, lack of restorative sleep, chronic stress, and sedentariness.

This important relationship between our day-to-day choices and the "set point" of inflammation, coupled with the role of inflammation as a causative factor in so many of our most feared medical conditions, is empowering, as it means that in a very real sense each of us can take control of our health destiny when we make good lifestyle choices—choices that ultimately bring us benefit through modulation of inflammation. And well beyond diseases, the actual process of aging itself is now being looked upon as a manifestation of ongoing inflammation, so much so that the term *inflammaging* has now entered the scientific lexicon.

Reining in chronic degenerative conditions and targeting the aging process are ideas that have appeal to those of us who are getting on in our years. That's for sure. But, more important, it is only very recently that scientific research has similarly linked inflammation to some of the most widespread and certainly challenging health issues facing a younger segment of our population, the adolescent. Mental health problems, including anxiety, depression, and other mood disorders, are rampant in school-age children, and the response for dealing with these issues from health-care providers is, more often than not, pharmaceutical. Were it not for their general lack of efficacy and significant side-effect profiles, drugs for kids might seem like a good option. But the most obvious flaw with this approach is that focusing on a pharmaceutical intervention in this context generally means ignoring the underlying cause or causes of the issue at hand.

Gaining an understanding of what underlies mental health issues in adolescents allows parents and health-care providers to focus on the fire, not just the smoke. And inflammation is a good place to start. In the pages that follow, Dr. Kenneth Bock takes us on a journey through a vast array of potential candidates that may be working individually or in concert to amplify the production of inflammatory chemicals in the bodies and brains of adolescent children. And whether addressing thyroid function, blood sugar, Lyme disease, food allergy, or even toxic exposure, Dr. Bock reveals that a final common pathway toward mental health disturbance that is shared by these seemingly disparate avenues is amplification of the inflammatory response. As such, *Brain Inflamed* explores *why* inflammation is enhanced and then reveals *how* to bring about its resolution with the goal of resolving the mental health issue that represents its manifestation.

But, most important, the following text imparts the compassion of a medical doctor's decades of a goal-focused endeavor to bring about solace for adolescents and their families who are challenged and looking for hope.

David Perlmutter, M.D., August 2020, aboard
Mystic Eagle, British Columbia, Canada

Introduction

I've seen a lot of complex cases over the thirty-five years I've been a practicing physician, twenty of which have been focused on children's health. It's enough that I've become known as a bit of an expert in helping people who might otherwise fall through the medical cracks. And over the last decade, the number of children who fit this description have grown more than I would have predicted. Increasingly, I have patients whose emotional volatility, low energy, and anxiety negatively affect their school performance, friendships, and even their cognition. Those are often my relatively mild cases. Others exhibit tics or signs of obsessive-compulsive disorder (OCD). By the time some kids get to me, they've often been diagnosed with a psychiatric disorder and been prescribed multiple psychotropic medications, from antianxiety drugs and antidepressants to stimulants and antipsychotics—and some of them aren't even in the fifth grade. I'm also seeing an uptick in severe cases of mood disorders—more persistent, more extreme, and often more harmful examples of the conditions above, like severe depression, bipolar disorder, and rapid cycling mood swings. These kids aren't sad and surly, they're literally pulling out clumps of their mother's hair. Their handwriting deteriorates. Their attention wanders. They isolate themselves and sometimes become self-injurious. Their parents come in aghast that their sweet, well-behaved children have transformed into raging, aggressive bullies or frightening, morose aliens, sometimes literally overnight.

I'm not the only one noticing a shift. Just read the headlines. Anxiety and depression have repeatedly been heralded as a national scourge of the tween and teen years and been featured on the covers of *Time* and the *New York Times Magazine*. While the diagnostic criteria for anxiety, depression, OCD, and post-traumatic stress disorder (PTSD) changed a little in the last update of the *Diagnostic and Statistical Manual of Mental Disorders* (*DSM-5*), the principal classification system used by psychiatrists, it wasn't enough to explain the increasing prevalence of these diagnosed mental health disorders in our tweens and teens. Yet a 2016 Child Mind Institute report announced that "One in five children suffers from a mental health or learning disorder, and 80 percent of chronic mental disorders begin in childhood." According to the National Institute of Mental Health, half of our adolescents have been diagnosed with a mental disorder, and of that half, about 20 percent will find it severely disruptive to their lives.

That's a lot of suffering children and parents seeking answers. But what if in the wake of what some are calling a mental health epidemic, we're looking for answers in the wrong place? What if it's not our children's minds that are sick, but something else?

In my practice, I dig deep to find that something else—if it exists—and I've frequently discovered that sometimes what looks like a purely psychiatric issue can in fact have biological underpinnings. For example, I've seen:

- OCD and tic disorders caused by post-strep infections
- Depression as a symptom of thyroid dysfunction
- Anxiety and fatigue as a result of adrenal dysfunction
- Mental sluggishness, cognitive dysfunction, and psychosis due to vitamin B-12 deficiency
- Behavior disorders, specifically agitation, connected to pinworm infection
- Oppositional defiant disorder (ODD), panic attacks, and rage episodes brought on by Lyme disease and Bartonella

- Hyperactivity, ADHD, conduct disorders, and mood disorders with tantrums provoked by allergies
- Depression, lethargy, separation anxiety, mood swings, paranoia, and schizophrenia precipitated by gluten sensitivity
- Frequent irritability and fatigue induced by reactive hypoglycemia (low blood sugar)
- Heavy metal (lead, mercury) toxicity displaying as ADHD, cognitive dysfunction, and emotional instability
- Infection-triggered autoimmune brain inflammation (ITABI) that has been mistaken for mood disorders, anxiety, separation anxiety, OCD, tic disorder, insomnia, depression, dysgraphia (impaired writing ability), dyscalculia (impaired math ability), nighttime incontinence, hallucinations, eating disorders, memory issues, and meltdowns.

In many of these cases, once I've figured out the underlying issues we're really dealing with and prioritized their treatment, I've been able to help these children regain their health, their stamina, and their confidence. When the symptoms have been especially severe or disruptive, it felt like I was healing the whole family, not just the child.

My goal was never to become a Sherlock Holmes for childhood neuropsychiatric problems that resist conventional treatment. The path to this book actually began during a two-day autism conference back in 1998. I had been a practicing physician for about fifteen years, and a year earlier had published my first book, *The Road to Immunity*. In the course of writing I had become interested in something called transfer factors, tiny peptides (small chains of amino acids) found in maternal colostrum—the nutrient-rich secretion produced by postpartum mothers before their breast milk comes in. Since it's well documented that colostrum helps support an infant's immunity, I was curious to learn whether these peptides could be used to help battle certain conditions, including allergies, autoimmune disorders, and autism, a diagnosis that was rapidly increasing around the turn of the millennium. Some of the coordinators of an autism conference heard about my research and invited me to speak. When I saw my time slot was the last hour of the final day, I assumed I'd be presenting to an empty room; most people would have either started to head home or convened at the hotel bar. To my surprise, when I took my place at the lectern, several hundred people stared back at me. In those days, autism conferences could draw a thousand attendees—parents, doctors, psychologists—all alarmed at the rising rates of childhood autism and searching for whatever information they could find that might help. I carefully explained my research, received some brisk applause, and went home unaware that the entire course of my career had just changed.

Almost overnight, I started getting phone calls from parents with autistic children. There just weren't yet many doctors who worked with autistic kids, especially integrative medicine physicians. I usually didn't either, but I was willing to try.

It was hard. There was not one thing that caused autism, and there was no cure, but through my integrative approach I was frequently able to ease symptoms and sometimes even launch major turnarounds. The community of parents of children with autism is enormous and tight knit. They connect with one another and share resources. When parents learned that autistic children in my practice were getting better in ways they hadn't gotten better before, I started seeing kids from all over the country, and eventually from all over the world.

With such a large pool of patients, it became clear to me that autism was not monolithic, but in fact could be thought of as "autisms." I began to recognize the importance of subtyping the children by the various underlying factors potentially contributing to their condition. One child's autism might be related to autoimmunity, another's might be exacerbated by a metabolic imbalance, and still another child might see improvement when we addressed his gluten sensitivity. I developed an expertise in identifying and ameliorating nutritional imbalances, toxicity, oxidative stress, immune system imbalances, and chronic inflammation.

In addition, as anxiety is a common comorbidity with autism, I made it a priority to find ways to ease these patients' nervous tension and agitation, closely following the most cutting-edge research and carefully translating it when possible into effective treatments. Many of my patients had siblings who, while not autistic, also exhibited symptoms of anxiety. When their parents saw that my methods helped ease the anxiety of their autistic children, they often asked me to treat their neurotypical kids for anxiety, too.

As the rates of anxiety continued to grow around the nation, and word got around that I'd had success in treating it, numerous parents without autistic children started bringing their kids to see me. These were tweens and teens who had inexplicably gone from happy and well adjusted to sullen, anxious, lethargic shadows of themselves, exhibiting symptoms from common anxiety and depression to uncontrollable tics and OCD, psychosis and unbridled rage or aggression. And more often than not, frequently in concert with mental health professionals, I was able to help them get better. My secret was an integrative medicine approach that helped me determine when a child is going through normal teenage angst, when he's truly in the grip of a psychological illness, and when his symptoms are in fact the by-product of an underlying infection, inflammation, autoimmune issue, or medical disorder, even when initial lab results might suggest otherwise.

As a clinician, I cannot ignore my observations and impressions in favor of lab results, especially when lab results can be flawed. If a child has symptoms that look like a thyroid disorder, sound like a thyroid disorder, and feel like a thyroid disorder, it's sound reasoning to consider attacking it like a thyroid disorder—despite the lab results not being frankly abnormal—to see if the treatment can produce improvement.

Relying on my deep medical knowledge in conjunction with instincts honed by thirty-five years of clinical experience has helped me identify treatment options with high-benefit, low-risk ratios, in turn helping thousands of kids get their lives back. The medicine I practice isn't a paint-by-numbers exercise in which each symptom gets plugged in to reveal a preconceived picture; I work freehand with a blank canvas until a picture emerges. I remain open to every possibility, including the notion that not all conditions that present as psychological disorders are actually psychological disorders. I believe this is why I am able to catch so many diagnoses that others may miss.

Sometimes a panic disorder really is a panic disorder. But there are kids out there who spend their days getting shuttled between primary care physicians, neurologists, psychologists, and psychiatrists, each one offering a different opinion, a different label, and a different drug that may or may not offer reprieve. Without a conclusive diagnosis, kids like these are relegated to a lifetime of prescriptions and therapy. And while they might get better, they never get well. In these cases, I've often found that a panic disorder is being caused by something no one has thought to investigate, like an immune system gone haywire, or low blood sugar, or an out-of-whack adrenal system. With proper treatment that addresses the source of the symptoms, these children's neuropsychiatric issues—the violent mood swings, brain fog, lethargy, anxiety, depression, tics, and OCD—can lessen and even disappear with significantly lower doses of antidepressants, antipsychotics, mood stabilizers, or other psychotropic medications than are often prescribed in these cases, and sometimes without any medications at all.

Everyone lives on a spectrum of physical and mental health, one that's far more expansive than what we've generally been led to believe. A diagnosis of mental illness can change the trajectory of a child's life. My goal is to help parents figure out what their child's mental health profile looks like, whether it's made up of garden-variety growing pains or full-blown disorders, so they can target the root causes and ensure an accurate and complete diagnosis, thereby getting them closer to finding the right treatment.

Not everyone can get to my office for a consult, and there aren't enough hours in a year for me to meet with the number of people asking for my help. That's why I wrote this book. If your child is in mental distress or has become a live-in stranger, and nothing you've tried seems to ease their pain, this book may offer hope. In these pages, you'll learn to look for clues that your tween's or teen's psych issues may not actually be psych issues, or at least not exclusively so. While it's entirely possible that your child is in fact suffering from a psychiatric disorder like depression or anxiety, it's also possible that an undetected infection, inflammation, or autoimmune disorder has triggered disruption in the brain or other organs, and *that* is what is causing your child's anxiety, depression, obsessions, and compulsions. Hitting the symptoms with mood stabilizers can improve a child's mental state but will do nothing to get rid of a possible infection or to quell an autoimmune reaction. I'll give you the tools you need to help ensure your child gets the proper care and treatment they need, whatever the final diagnosis, as well as the understanding, resources, and support you've likely been searching for.

One of those tools is an awareness of the inherent social and emotional challenges of working with kids of this age. It can be tricky to advocate for kids who might simultaneously want to be cared for and left alone. The kids' buy-in is often a crucial component to their healing regimen, and we'll discuss how to talk to teens or tweens in a positive way that can help them feel empowered and willing to take control of their health and their life.

Many people come to me because they like that I practice integrative medicine, which is known to incorporate natural remedies and diverse modalities into treatment plans. Some are then shocked or even resistant when I prescribe or recommend a synthetic medication, like a low dose of an antipsychotic, an antibiotic, or a mood stabilizer. My job is to heal and protect, no matter what. Sometimes a helpful treatment has an unfortunate side effect, and together my patient or my patient's parents and I have to weigh the benefit versus the risk. If a child undeniably needs the treatment, I'll do my best to offset those side effects. For example, if I prescribe an antibiotic, I also have to protect my patient from potential side effects like dysbiosis (a disruption of the microbial balance in the gut, a condition that we will explore extensively in this book), so I might suggest a regimen of probiotics and prebiotics, in addition to certain herbs to help protect the liver. The true essence of integrative medicine lies in being open to various approaches and causative factors. A change in diet, supplements, and an anti-inflammatory medication can work together; they are not mutually exclusive. I'd never see such high success rates if I weren't able to use both pharmacological and natural approaches to treat complex illnesses.

I have no intention of pathologizing tweendom and the teens. These are special years that come with many challenges that can turn a mom's or dad's hair prematurely gray; but it's all part of a natural process of separation and establishing independence that heralds many opportunities for growth, and not just for kids. This book simply offers options and alternatives for kids who aren't responding to traditional treatments or therapy.

Without question, neuropsychiatric disorders like anxiety, depression, schizophrenia, and OCD exist, and we're lucky to live in a time when we have access to excellent psych medicines and behavioral therapies that can bring relief to suffering children and their families. But it would be a tragedy for a child to be consigned to a lifetime of therapy and psych meds when a change in diet, an antibiotic, or even a tonsillectomy could eradicate the cause of his symptoms, possibly for the long term, and possibly for good. This book will help ensure that you know what to look for, what questions to ask, and what avenues to pursue so you get an in-depth evaluation and become the best medical advocate possible for your tween or teen.

Brain Inflamed introduces a concept I call the Mood Dysregulation Spectrum, which I developed to help illustrate the vast range of psychological symptoms experienced by tweens and teens, and the particular combinations that can suggest actual illness instead of more typical and expected growing pains, identity exploration, or rebellion. We'll do a thorough overview of the immune system, as well as of the bidirectional nature of the gut-brain-immune axis and what can happen when something disturbs its delicate balance. A significant body of research indicates that inflammation—including gut, brain, and immune inflammation—caused by these imbalances is at the root of all kind of ailments, including young people's mental health disorders. These lessons will be key to understanding how physical ailments can cause psychological symptoms.

With this scientific grounding, you'll be prepared for a chapter-by-chapter exploration of the medical and biological dysfunctions that can cause psychological illnesses in young people. Inflammation is just one such dysfunction; we'll also explore infections, autoimmunity, and metabolic and hormonal imbalances including hypothyroidism, adrenal dysfunction, and reactive hypoglycemia. To illustrate how symptoms can manifest in real life, I'll share the stories of real patients, once lost in a haze of pain and confusion, and their parents, who stopped at nothing to help them find relief. You'll read about treatment options and find questions at the end of each chapter to help guide your thinking as you consider whether your child could be afflicted with any of these dysfunctions or imbalances. Throughout, we'll cover preventive strategies.

One thing I've learned over my years of speaking with groups of parents at autism conferences is that it's important to meet people where they are. At each speaking event, I'd find newcomers who were just starting to learn about the disorder, while others in the audience had been researching autism for years and were eager for more detailed, advanced information from professionals on the front lines. While this book was written to be accessible to anyone, occasionally I have included some in-depth detail into cellular or hormonal mechanisms. These might seem excessive or overwhelming to someone new to the topic, but they're necessary for anyone interested in taking a deeper dive into the subject matter. It's completely okay, however, if you prefer to quickly scan or even skip it. Don't stress! What's important isn't that you remember the details, but that you gain greater awareness of inflammation's role in adolescent anxiety and mood disorders, and the numerous paths you can pursue to investigate and treat them.

Though I've been fortunate to be able to solve many medical mysteries over the course of my career, I do want to caution that this book isn't a cure-all. Nor would I suggest that parents can or should diagnose or treat their children themselves. What I can promise is to inform you about the possibilities, controversies, and treatment options available. Use this book to initiate the preliminary detective work crucial to fast-tracking your child toward the appropriate evaluation and treatment. My goal is to arm you with the important questions to ask when seeking treatment from your own trusted medical practitioners, as well as instructions for what to do if you're not satisfied with the answers you receive.

Some readers may conclude that their tween's or teen's behavior and moods are linked to purely psychosocial causes—unrequited love, academic stress, parents' divorce, online bullying, or any other number of the typical problems today's young people face as they navigate the challenges of growing up. But a parent knows. If your child isn't acting like herself, if her life is changing for the worse, if she is missing school, or if something just doesn't seem right and you're scared, I hope this book provides help. Somewhere on the spectrum between perfectly normal teen angst and mental illness is a range of symptoms that are often surprisingly interconnected and surprisingly treatable, sometimes even with natural approaches instead of pharmacologic ones. *Brain Inflamed* will reveal how to examine and address these issues in a coherent, effective way, offering realistic hope to any parent seeking direction, answers, and peace of mind.

This revolutionary approach to mental health has helped countless kids heal, reclaim their joy, and fulfill their potential. It just might be able to help your child, too.

Brain Inflamed

CHAPTER I

Welcome to the Spectrum

Carol and her eleven-year-old son, Sean, were cleaning up the kitchen together when she first noticed the hops. Initially she thought he was bopping along to the hip-hop he'd chosen to play over the speakers, but after the fourth track she'd had enough and ordered Alexa to play Miles Davis instead—definitely not music her son was going to want to dance to. Yet as she scrubbed a skillet in the sink, she could feel the short little hops continue next to her, as though every minute or so Sean was bouncing on an imaginary pogo stick. "What's with the hopping?" she asked, handing Sean the sopping wet pan to wipe down with a dish towel.

"What hopping?" Sean asked.

As the days went on, the hops became more pronounced. Any time Sean was standing still, within a few seconds he'd pop up and down like a little life buoy. But when Carol asked him about it, he looked at her blankly. What hopping?

A week or so later, Sean started to tap. Walls, ledges, the tops of chairs, his nose. His glass against the table. His pencil against his notebook. And every tap made with one hand or elbow or thumb had to be repeated with the other hand, elbow, or thumb. Carol found it mildly irritating, but it didn't seem to bother Sean. Then one

day she got a call from the principal's office at Sean's school, asking
her to come pick him up. He'd instigated a fight and had received a
two-day suspension. Carol was too surprised to be angry. Sean was a
mild-mannered boy and had never been disciplined at school before.

His face was downcast and glum as she walked him out of school,
and neither said much during the drive home. That evening over dinner, Carol finally asked Sean to tell her what happened. He explained
that he hadn't instigated a fight at all. Cameron had been standing
in front of him in the lunch line, and he'd pressed his hand into the
middle of Cameron's back, and Cameron had shoved him, so he'd
shoved back.

Carol frowned. "Sean, that doesn't make sense. Why would
Cameron shove you for that?"

Sean couldn't look at her. "Well, he asked me to stop, but I couldn't."

"What do you mean? How many times did he ask you to stop?"

Sean shrugged. "I dunno. Bunch of times."

Carol was starting to get annoyed. "Sean! Why wouldn't you stop
touching someone when they asked you to? That's little kid behavior.
You know to keep your hands to yourself!"

Sean's face crumpled as his eyes blurred with tears. "I didn't say I
wouldn't stop. I said I *couldn't*."

Mark poked his head into his daughter's room. Everything looked
the same. There were the piles of clean clothes sitting next to the
dresser instead of inside the drawers where they belonged. A stack
of books lay next to the bed. Taylor Swift pouted and Pink snarled
from their respective posters on one wall, while a collection of bright
yellow emojis decorated another. In the middle of her rumpled bed
lay Mark's thirteen-year-old daughter, Rima. Mark had never been
able to sleep if there was even a crack of light in his room, but the
midafternoon sun didn't seem to bother Rima at all. It never did.
Mark thought about all the other parents who swore to him that
their teenagers slept ungodly hours, too. But in the middle of the
day? And when their teens were awake, did they come downstairs

every now and then? Mark felt like he hadn't had a real conversation with Rima in months. Even worse, over the course of a few days it seemed like her personality had changed dramatically. Once chatty and open, she was now frequently irritable and mostly communicated in grunts and sighs, interacting with her parents and her younger brother only when necessary. Mark remembered being a teenager and wanting to keep to himself, too, but he also remembered going to the movies and playing sports. Rima had pulled back from extra-curricular activities. She'd missed so many soccer practices because of stomachaches that Mark finally pulled her from the team. Yet the pediatrician couldn't find anything wrong with her. Even a colonos-copy had revealed nothing out of the ordinary.

Mark knew girls were different from boys, but *this* different? His wife, Mariam, also admitted that she was finding Rima difficult to figure out. Mariam had holed up in her room at that age, too, but she'd been on the phone, talking for hours with friends while twirling the long, curly phone cord around her fingers. Rima wasn't actually talking to anyone, as far as her parents could tell. Her entire social life seemed to be happening through Snapchat, Instagram, and TikTok, which she scrolled through mindlessly, her eyes dull through her un-kempt bangs.

Mark moved a little closer to look at his daughter's sleeping form, hoping to see in her softened face a trace of the happy little girl she'd been not so long ago.

April sat at her desk in her home office, the door closed, her face in her hands, waiting. Elsewhere in the house, her son, Nicholas, stormed like a hurricane.

The day had started off well enough. It had taken him forever after she'd heard his alarm clock ring that morning, but Nicholas had finally made it down the stairs looking much like his old affable self. He'd thrown a sack lunch together, kissed his mother goodbye, and headed out the door for school. April, who worked from home for an insurance company, had settled down at her desk for a productive day. She'd been surprised when she'd heard the back door open, announcing her son's return home. How had the day flown by so fast? She headed into the kitchen, hoping the offer of a homemade muffin might bribe Nicholas to spend a few minutes with her and tell her about his day before he disappeared into his room to do homework. He was pouring himself a soda when she walked in, and as soon as she caught his eye, she could tell his mood had changed from the morning. There was something hard in his gaze that wasn't there before. Taking a breath, April decided to try to engage anyway.

"Hey, honey!" she said with a brightness she didn't feel. "How was your day?"

"Fine." Nicholas put the soda bottle back in the refrigerator and grabbed his glass, trying to move past her.

"We still have some of those muffins I made over the weekend. You want one?"

Nicholas paused. "The blueberry ones?"

Aha! She'd caught his interest. "Yep!" she said as she swung open the refrigerator door and rummaged around, looking for the right Tupperware container. She pulled it out, took the lid off the top, and held it out to her boy. He looked into the container and frowned.

"Those aren't blueberry."

She pulled the muffins closer to take a better look. Shoot, he was right. They must have finished all the blueberry muffins on Sunday.

"Well, that's okay, isn't it?" she asked hopefully, extending the container back toward him. "You like the banana ones, too, don't you?"

Nicholas pressed his mouth together for a minute, as though he were thinking of what to say. Without warning, he grabbed the Tupperware and hurled it across the room.

The next few minutes were a blur as April tried to calm her son down, but his rage was too great as he shoved past her and slammed his backpack onto the kitchen table so hard the overhead lamp swayed a little. His breathing sounded jagged, like a panicked animal. April retreated to her office. Her hands shook as she texted her husband.

"He's gone crazy again. Come home now. NOW!"

The Teen State of Mind

Every teenager can be moody. What parent hasn't joked about holding on for dear life through the hormonal surges, impulsive acts, rebellion, and surliness that turn life with a teen, or even a tween, into the scariest roller coaster ride they've ever taken? Teenagitis is often easy to diagnose: if your teen is Severus Snape at home, but they still have friends, and those friends' parents tell you what a pleasure your child is to have around, that's teenagitis. That's a kid who can turn behavior on and off at will, while likely going through the normal teenage rite of passage, testing boundaries, expanding horizons, claiming independence, and separating themselves from parents who on their thirteenth birthday magically became the dumbest people on the planet. But then there are the other kids, the ones gripped by uncontrollable mood swings, tics, compulsions, lethargy, or rage. And many of them may be as scared and sad about their transformation as you are.

Your child may not be exhibiting the identical symptoms of these children, but if you've picked up this book, it's likely you're feeling much of the same fear, frustration, confusion, and worry as their parents. First off: You are not alone. Not by a long shot. By the time they turn eighteen, as many as 50 percent of all children and teens will meet the diagnostic criteria for at least one mental health disorder. Five times as many high school and college students today say they are dealing with anxiety and other mental health disorders as young people of the same age did when surveyed during the Great Depression. One in five children are reported to have a mental health disorder, making it the most common health issue experienced by school-age children, anxiety—which can manifest as panic attacks, separation anxiety, or phobias, among other symptoms—being the most prevalent diagnosis, affecting as many as one third of adolescents. All those children represent millions of concerned, worried, and frequently frightened parents. Parents who would do anything within their power to ease their children's pain. Parents just like you.

No one asks to join this club, and yet the number of its members keeps rising. What could cause such a tremendous surge in teen and tween mental health problems, something not seen even in the generation that grew up during the Great Depression, many of whom never stopped stockpiling boxes of powdered milk and hiding cash under their mattresses?

Certainly it's possible that the rates aren't as different as we believe, that doctors simply have more advanced diagnostic tools at their disposal, and that we are merely more aware of these issues than we were before. I don't discount the possibility that awareness may be a small contributing factor, but evidence supports several other theories. One is that our diet is affecting our children's mental health. Preliminary research correlates a healthy Mediterranean-type diet high in vegetables, fruits, whole grains, olive oil, and (small, low-mercury) fish with lower risk for depression, while a diet high in sweets, refined carbs, processed foods, and high saturated fats—your typical Western diet, especially in busy or low-income families—is associated with higher risk for depression.

Another theory is that time spent in nature and free play both show strong benefits to mental health, yet today's children get far less of both than previous generations. Only about one out of four five-to-ten-year-olds get one hour or more of physical activity per day, and many teenagers don't get any more exercise than a typical sixty-year-old. Poverty, too, is a mental health risk. Children under the age of eighteen represent 33 percent of our country's poor, and in one study, youth living in two economically disadvantaged neighborhoods (one in Chicago, one in Portland, Oregon), with high rates of crime, unemployment, housing and food insecurity, and single-parent homes, showed up to seven times more long-term anxiety than their peers in the general population. Yet kids from higher-income environments face high rates of depression and anxiety as well.

Some researchers blame distant relationships between children and parents, or an uncompromising high-pressure culture of achievement. The time frame during which kids are allowed to be kids continues to shrink, eroded by an across-the-board cultural shift toward intensive parenting and higher academic expectations. Even their play is frequently high pressure, emphasizing tournaments, performances, and competitions.

When kids are not practicing or studying, they're steeped in an intensely polarized political climate, as well as very real concerns about their safety. In 2018, Gen Z teens reported feeling highly concerned about mass gun violence, sexual harassment, and family separations, among other issues, even more than adults, according to the American Psychological Association's report *Stress in America*. As Philip Kendall, director of the Child and Adolescent Anxiety Disorders Clinic at Temple University, observed that same year, they are "growing up in an environment of volatility, where schools have lockdowns. . . . We used to have high confidence in our environment—now we have an environment that anticipates catastrophe." And that was before the COVID-19 pandemic officially reached the U.S. in January 2020, closing schools and businesses; cratering some families' finances; scuttling graduations, internships, summer jobs, and camps; and forcing families to quarantine in order to avoid spreading a novel coronavirus that at the time of this writing has killed more than 250,000 Americans and sickened millions. Like wars, 9/11, or recessions, this event will surely have an impact on our young generation's perspective and sense of security. With today's adolescents already reporting increased feelings of loneliness before the pandemic, some psychologists, like teen mental health expert Jean Twenge, suspect the crisis could also have a deleterious effect on their mental health.

Finally, it would be impossible to discount the role social media has played in the higher rates of mental health illness in children. Twitter launched in 2006, the same year Facebook lowered its registration age to thirteen. Between 2007 and 2012, the National Survey of Children's Health found that anxiety diagnoses jumped 20 percent in children ages six to seventeen. Fifty percent more teenagers were clinically diagnosed with depression in 2015 than in 2011, right around the time social media became ubiquitous in the teen population. Smartphone adoption in the United States surpassed 50 percent in 2012. Since then, many kids enter their early teen years feeling obligated to put themselves on constant display, in return basking in a never-ending glare of marketing, FOMO, and peer pressure. Says Janis Whitlock, director of the Cornell Research Program on Self-Injury and Recovery, "If you wanted to create an environment to churn out really angsty people, we've done it. . . . They're in a cauldron of stimulus they can't get away from, or don't want to get away from, or don't know how to get away from."

In sum, more research needs to be done to determine whether the increase in children struggling with anxiety and depression is because of poor diets, pressure to perform, a sense of instability, social media, or most likely, as we'll discuss in the next chapter, a combination of all these factors. The one thing researchers do seem to agree on is that a significant number of our kids are stressed in a way children in this country have never been before. The question is, with this much stress poisoning our children's lives, why isn't every kid turning inward, or transforming from a Jekyll to a Hyde?

The Mood Dysregulation Spectrum

To fully address the needs of children suffering from neuropsychiatric problems, we have to be open to the possibility that any symptom could be caused by a complex interaction of culprits, including medical, neurologic, psychiatric, gastrointestinal, environmental, genetic, metabolic, hormonal, nutritional, autoimmune, and infectious. One child could present the most obvious manifestations of panic disorder and OCD, another might be acting depressed, yet another could be experiencing episodes of inexplicable rage and defiance, and all their symptoms and behaviors could be caused by an underlying strep infection that has sparked immune dysregulation and inflammation of the brain. Or they might be suffering from an imbalanced flora in the gut, or a tick-borne disease like Lyme. Similar to strep, the latter can cause an infection as well as immune dysregulation and inflammation of the brain that can look like panic disorder, OCD, ODD, depression, or intermittent explosive disorder. Again, whether children who contract strep or Lyme disease exhibit any of these symptoms, or none at all, has everything to do with their unique combination of genetics, the level of exposure, the timing of their exposure, and their overall immune health.

The wide range of moods and symptoms I see in my neurotypical patients are as diverse and extreme as those exhibited by my autistic patients, where high-functioning children with Asperger's can thrive in mainstream educational settings, while others with low-functioning autism might be mute, uncommunicative, aggressive, or exhibit self-stimulatory behavior (stimming), like flapping their hands or rocking. And yet while the symptoms of autism can vary greatly in their phenotypes (observable physical and/or behavioral presentations), they all fall under the same catchall diagnosis: autism spectrum disorder. Over the years it became clear to me that the symptoms of thousands of neurotypical children could be placed on a similar spectrum. I call it the MDS, or Mood Dysregulation Spectrum. It's a useful concept that helps people think of their children's symptoms in a new way, as part of a much more expansive paradigm than the one that merely limits us to a psych diagnosis. Being aware that there are multiple medical conditions that could contribute to your child's condition is the first step toward getting them better. Once you can identify your child's unique MDS profile, you can start to decode what could be causing their symptoms.

The list of disorders and their symptoms can range from mild to moderate to severe. As a baseline, ideally we'd like for every child's MDS profile to look something like this:

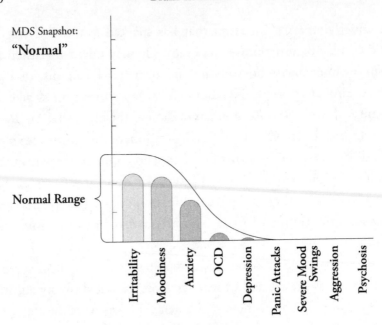

MDS Snapshot:
"Normal"

Normal Range

Irritability Moodiness Anxiety OCD Depression Panic Attacks Severe Mood Swings Aggression Psychosis

But a child coping with the typical psychosocial issues of their age—keeping up in school, chafing against rules, maneuvering the landmines of relationships and social media, becoming aware of their place in the global world and its sometimes frightening complications and contradictions—might look like this:

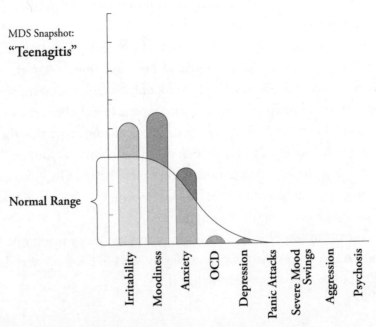

MDS Snapshot:
"Teenagitis"

Normal Range

Irritability Moodiness Anxiety OCD Depression Panic Attacks Severe Mood Swings Aggression Psychosis

That's normal, too. That's a case of teenagitis.

But the graphs of symptoms of the children like the ones described in the first pages of this chapter might look like this:

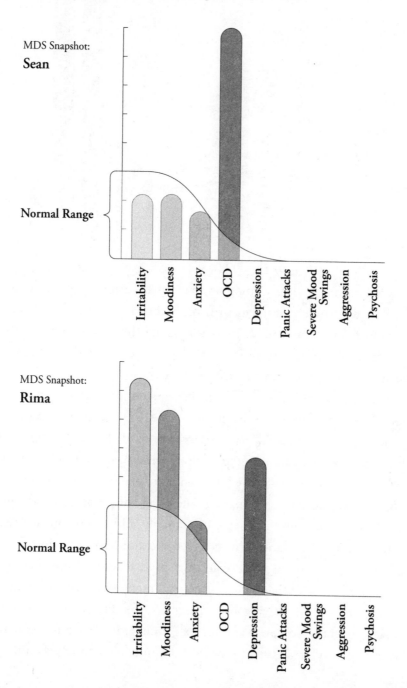

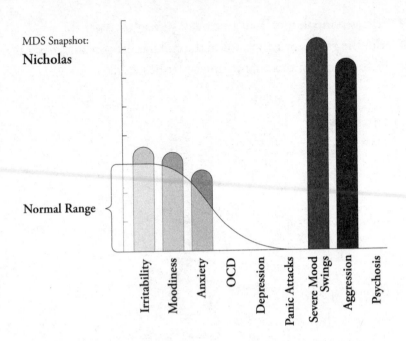

MDS Snapshot:
Nicholas

Normal Range

Irritability · Moodiness · Anxiety · OCD · Depression · Panic Attacks · Severe Mood Swings · Aggression · Psychosis

Something is clearly going on, but what? Compare the children's graphs to a baseline template of Lyme disease or strep-induced brain inflammation symptoms (while recognizing there's variability in the presentations of each disorder, as well as in each individual tween or teen) and you might start to see similarities.

Graphing a child's symptom profile and comparing it to an MDS template won't get you a diagnosis—and that's not its purpose. Instead, the MDS is useful for frustrated or worried parents who keep hitting dead ends, as these graphs, along with the questions following them, can offer some worthwhile avenues to explore and questions to ask. Every child's MDS signature will be as unique as a fingerprint, but as in fingerprint analysis, with enough "hits," you can increase the certainty of making a correct identification. You can download a blank template to use with your own child at braininflamed.com.

For all their individuality, of course, many children, no matter how much they're driving their parents crazy, fall squarely within what's generally accepted as "normal" teenagitis. The only remedy parents can offer are clearly defined boundaries and expectations, and unlimited stamina and love (for both them and for yourself) as you wait for them to grow out of it. However, certain spikes and combinations of symptoms create signatures that can be mapped to physical illnesses and disorders that manifest psychologically, neurologically, and behaviorally. These are the cases we'll explore in this book. Should you recognize your child in any of them, you'll have an expanded understanding of the symptoms' possible causes as well as potential treatments to discuss with your health-care team.

A Family Affair

Before we go further, I want to acknowledge that coping with a child's mental illness is a family affair. I hope that as you seek out help for your child, you'll also seek out help for yourself and other involved family members, even if it turns out your child is just dealing with one heck of a prolonged case of angst or moodiness. When one member of the family unit suffers, often everyone does. We're not only traumatized by threats to ourselves; witnessing threats to the ones we love, even in the form of illness, can register as traumatic experience as well. Many of the parents of my sickest patients will admit they feel traumatized by their illness, with accompanying symptoms of post-traumatic stress disorder (PTSD) such as anxiety, hypervigilance, or an inability to feel safe. Parents worrying about their children—especially those who have been severely affected—often need a professional to help them regain their confidence and equilibrium, as well as help them make sure the fabric of their relationships remains intact.

The integrity of the family unit and its vulnerability when one of its members is hurting isn't much different from the way the systems of the body interact and depend upon one another. The immune system, the gut, the brain—they're all connected. Healing any of them requires a deep understanding about their pivotal role in our health, and the effects that inflammation and imbalances can have on our mind and body. So that's where we will start.

QUESTIONS TO CONSIDER:

When did you first notice your child's symptoms?

Was the onset of the symptoms sudden or gradual?

Was there any event or illness that might have preceded and/or precipitated these symptoms?

Are their symptoms cyclic or protracted/prolonged? If cyclic, do their moods cycle rapidly, from one to another?

Do they behave very differently at home than in school or at their friend's house?

Are they staying home from school and/or not hanging out with their friends?

Do they seem nervous, hyper, or agitated?

Do they have trouble falling or staying asleep, or conversely are they sleeping too much?

Do they frequently seem sad?

CHAPTER 2

Immune System 101

Many Americans received a free crash course in immunology during the early days of the COVID-19 pandemic as they tried to learn more about the novel virus, understand their risk, and become conscious of the best precautions to take to avoid infection or infecting others. Yet for all their heightened interest in and appreciation of their immune system—the body's defense network against illness and infection—most people still underestimate how much it factors into almost every aspect of their biological and psychiatric function. Before we delve into the individual issues that may be contributing to your child's illness, it's important you have an understanding of what the immune system does, how it works, and the intimate ways in which it's connected to other bodily systems. Ultimately, this information will also help reveal how inflammation could start in one part of the body, like the gut, and wreak havoc on another, like the brain. It also helps us understand our options when trying to stop it. In journalist Susannah Cahalan's bestselling memoir, *Brain on Fire*, the doctor who finally diagnosed her acute and sudden psychosis compared the brain inflammation that triggered her near-fatal illness to a fire. As in any fire, you'd need to know what was causing the flames in order to successfully extinguish them. Was the source electrical? Thermal? Chemical? Each type of fire requires a different method for putting it out. Water on an electrical fire, for example, only makes things worse. And you'd also want to know how to prevent the fire from reigniting.

Our knowledge of the immune system is constantly evolving. It can seem a bit complicated, but I've been presenting a simplified model of the immune system to parents for years, and over time many have gone out of their way to tell me how helpful it was to have this context for understanding their child's illness. I hope you will feel similarly.

What Affects Our Immune System?

There are two imperatives to good health: flexibility and resiliency. I'm referring not just to character traits, but to the factors that foster health at the cellular level. Their absence or presence explains how people can be subjected to the exact same stresses yet respond completely differently. After all, though almost every child growing up in America today will have some exposure to fast food and Instagram, not every one of them or even the majority will become clinically depressed, anxious, or ill.

I like to use the image of a kettle to explain how environmental or genetic elements can affect our immune system's flexibility and resiliency, thereby influencing its ability to adequately protect and defend us from stress. From the moment we're born, and in some instances even earlier, that kettle starts to fill—the effects of our environment, diet, relationships, and lifestyles layering themselves over our baseline genetic predispositions. The higher the contents of our immune kettle rise, the closer we get to overflow, and the more likely our minds and bodies will start to experience symptoms.

First Layer: Genetic Predispositions

From conception, every human being is a unique world unto themselves, their DNA influencing everything from their lifelong risk of

Immune Kettle

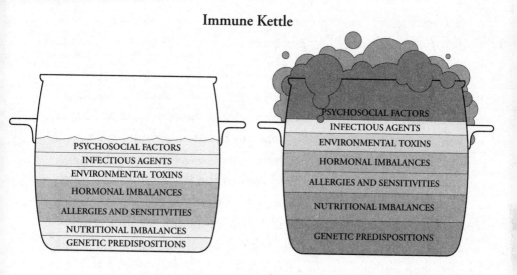

disease to their tastes and preferences to the shape of their developing body. Those genetic predispositions make up the first layer of our immune kettle. Luck of the draw will determine the height of this layer and therefore how well or poorly your body can tolerate the addition of other layers. Some people are born with a sturdy immune system that takes up little space in the kettle, while others will find their kettle quickly starts to fill with susceptibility to viruses and infections, trouble regulating their blood sugar in between meals, or allergies and autoimmune reactions. Though many of us inherit a genetic predisposition to any number of conditions—such as asthma, food allergies, or mood disorders—they often won't manifest unless we come into contact with corresponding environmental or dietary triggers. And even then, we might not react unless additional layers of our kettle rise higher than our immune systems can handle, causing the whole thing to boil over. For example, you could have a genetic predisposition to celiac disease, an autoimmune disorder we'll study further in chapter 8, and never experience any symptoms without the accompanying presence of an increased permeability of the small intestine *and* the dietary trigger gluten.

Second Layer: Nutritional Deficiencies, Insufficiencies, and Dependencies

Every single thing we put in our mouths, from a spinach salad to a Pop-Tart, gets broken down into fuel for our bodies' cell and organ function. Too much food and refined carbohydrates puts us at risk of obesity, inflammation, higher cholesterol, insulin resistance, diabetes, and high blood pressure, among other ailments. Excessive amounts of certain healthy nutrients can also be harmful—too much protein, for example, can cause constipation; too much calcium can cause kidney stones. But when we give our bodies too little food, or too much junk food with little nutritional value, we not only starve our brain and muscles of energy and strength but we also rob our immune system of the nutrients it needs to help run its sophisticated defense and repair mechanisms. For most people, these deficiencies are a result of consuming excessive amounts of processed food—especially starchy and sugary snacks that overload our system with glucose—while not eating enough whole foods that provide essential vitamins and nutrients. Some deficiencies are also linked to genetics, as is the case with inherited metabolic disorders (such as lactose intolerance or gluten intolerance) that can hamper our ability to digest certain foods.

Many of my patients aren't nutritionally deficient, but nutritionally *insufficient*. For example, your child's lab tests may reveal a low-normal level of vitamin D, which is a crucial vitamin for a robust immune system. But I run metabolic, biochemical, and nutritional evaluations that give me a more detailed picture of your child's health. Those tests might reveal that a vitamin D level that is low-normal to the general healthy population may be insufficient for someone with their particular genetic makeup living in their particular environment; that insufficiency may in fact be contributing to increased autoimmunity, in which a person's immune system mounts an offensive attack against their own tissues and cells, mistaking them for outsiders (nonself), instead of mounting a defensive attack against external or foreign threats such as viruses, bacteria, or cancer cells (more on autoimmunity later in this chapter). I've seen people with immune dysregulation whose standard lab tests for certain nutrients technically registered as "normal," yet enhanced testing of those nutrients showed they were actually insufficient. I treated them with additional nutrient support and subsequently saw improvements in their immune function. Furthermore, I'm not only paying attention to the amount of nutrients, but also to the relationship of one nutrient to another. For example, if I have a patient showing signs of bipolar disorder and I conduct a test that shows elevated levels of kynurenic acid, I also learn the patient probably has a functional vitamin B-6 deficiency that needs to be corrected. Why? Because kynurenic acid can't be metabolized without the active form of vitamin B-6; if there is more kynurenic acid than there should be in the urine, it's because there's not enough vitamin B-6 to break it down. And vitamin B-6 insufficiency can cause psych symptoms such as severe mood swings or delusions, which can be indicative of bipolar disorder or schizophrenia.

This same testing usually reveals that most of us have one, two, or three nutritional dependencies, meaning each individual body (and brain) needs a higher level of certain nutrients to function at its best. For example, you might need more zinc than I do. I may require more vitamin B-2 for optimal functioning than you do. Dependencies and insufficiencies are extremely specific to each individual, a concept referred to as biochemical individuality. An immune system running at normal levels is great; an immune system running at optimal levels, possible only when the needs of your biochemical individuality are being met, is even better—and can feel better, too.

Third Layer: Allergies and Sensitivities

There are people who grow up knowing that April showers bring May flowers as well as blooming trees, pollen, and utter misery in the form of sniffles, sneezing, coughing, and itchy eyes. Others only discover they have allergies when they adopt a cat, eat a peanut, or get exposed to certain chemical pollutants (although the latter is usually more of a sensitivity than an actual allergy). Most people accept their allergies as unavoidable and manage them to the best of their ability through the use of antihistamines or avoidance of the offending substance. But I've learned that sometimes when you examine the totality of a person's immune kettle and address deficiencies, imbalances, and toxic loads so that the immune system can function properly, you can minimize and perhaps even eliminate the manifestation of the allergy. You can do the same for sensitivities—which can be delayed and provoke less severe reactions than classic allergies, such as joint pain or headaches as opposed to hives or breathing problems—and intolerances, which can be caused by a deficiency in the enzyme necessary to break down certain foods and can tend to manifest as unpleasant digestive issues. Fine-tuning the immune system gives it more leeway to protect itself against offending substances.

Almost every day I meet patients who first came to me for help with one issue, say GI symptoms, only to report when they come back for follow-ups that not only are their intestinal problems gone, but so are their chronic headaches, fatigue, brain fog, depression, or anxiety. That's what happens when you give your body what it needs *and* remove what is harming it. For some, it's the first time in years they've known what it's like to feel vibrant, energetic, clearheaded, and calm. Witnessing these kinds of transformations makes it a joy to go to work every day.

Fourth Layer: Hormonal Imbalances

The adrenal and thyroid glands are part of the endocrine system, which produces and secretes hormones that regulate a wide range of essential bodily functions, including metabolism. Because the adrenals and thyroid produce hormones that literally affect every single cell, disorders can cause symptoms that range across the board:

obesity, loss of appetite, constipation, diarrhea, dry hair, hair loss, muscle weakness, low muscle tone, scaly skin, mental sluggishness, fatigue, depression, low blood pressure, hypoglycemia, and excessive sweating. Here, too, imbalances are often overlooked because what are considered normal hormonal levels are not always adequate for every individual. You can give someone with adrenal or thyroid disorders all the antidepressants in the world, but until you regulate and balance their hormones, they won't get fully better.

Fifth Layer: Environmental Toxins

Until the early 1990s, U.S. environmental policy was determined by analyzing data that focused on the assessed risks posed to the "average adult." Then in 1993, a report citing multiple studies established that children are more susceptible than adults to chemicals and toxins in their environment. This realization led to many legislative and regulatory reforms, in particular with regard to pesticides and pharmaceuticals. Unfortunately, while regulations have eliminated or at least decreased the obvious poisons in our environment, we are still inundated with pollutants, plastics, industrial chemicals, heavy metals, and other toxins every day. They show up in the air we breathe, the soil in which we grow our food, the water we drink, and the animals we eat, in addition to our construction materials, furniture, bedding, clothing, and cleaning supplies. The Environmental Protection Agency (EPA) lists more than 86,000 chemical substances manufactured or processed (including imports) in the United States. We don't know exactly how many are harmful; most have never been thoroughly safety-tested. But a growing body of research performed by many scientists and medical practitioners—including me—offers damning evidence that many of the foods and products in our homes are causing harm, especially to our children. This layer is difficult to measure, but it is higher than it should be for just about everyone.

Sixth Layer: Infectious Agents

When I first came up with this immune kettle concept, I was still unaware of how frequently the conditions I treat were caused by multiple hidden, low-grade, simultaneous infections. These infections often don't present in any dramatic fashion; many people don't even know they're carrying them. But when they linger and become chronic, infections can burden the immune system and weaken its defenses, so that when something else comes along that needs its attention, it is less able to respond as needed. Strep, for example, can hide in your tonsils for years without ever causing painful strep throat. Epstein-Barr, a common virus that can cause mononucleosis, lies dormant in the body for the rest of your life even after you no longer have the acute illness. If you contract Lyme disease, however, the Epstein-Barr virus can reactivate because the immune system now has to divert some of its attention to the Lyme instead of keeping the Epstein-Barr in check. It's just like in a war—there are a finite number of defensive resources, and if you've concentrated your arsenal to protect one entry point, others may become more vulnerable. These lingering viruses and bacteria may not be extremely aggressive in and of themselves, which is why they're not your immune system's priority. Yet they can still do damage, and if your system is already weak, their presence can raise your body's "boiling point" and cause your kettle to overflow, with damage cascading in all directions.

Seventh Layer: Psychosocial Factors

There are many new pressures that weigh on tweens and teens today: social media that discourages face-to-face interaction, encourages aggression and bullying behaviors, promotes social comparison (in which we determine our worth based on what we think others have), and leads users to feel isolated and sad; high parental and academic expectations; the awareness that rapidly shifting technology will change how the world works by the time they enter adulthood; a perception of environmental instability and danger, likely made more acute by the COVID-19 pandemic; and media that sets unhealthy and unrealistic body-image standards. All that stress acts similarly to a toxicant, with similar biochemical and metabolic perturbations. When we're in a state of stress, depression, or grief, our adrenals release the steroid cortisol into our system to help our bodies cope. However, steroids suppress the ability of our white blood cells to fight infection. Children experiencing a sustained aberrant physiological response due to constant stress or negativity can suffer immune impairments that may be long lasting. As researchers in the burgeoning field of psychoneuroimmunology increasingly reveal, our mental health has a tremendous effect on our overall health.

Inflammation

Inflammation—the immune system's defense response to cellular injury, pathogens (i.e., a virus or bacteria), or other cellular harm—isn't one of the layers of the kettle, but its effects can span across most of them. Sometimes it is the cause of our symptoms, sometimes it is the symptom itself, and sometimes it can be both. For more than a decade, medical researchers have suspected that inflammation is a major culprit in almost everything that ails us, and new research continues to link inflammation to a wide range of illnesses, from heart disease to cancer to diabetes to neurogenerative disease including Parkinson's and Alzheimer's as well as psychiatric disorders. The brain and the immune system are in constant communication, and we now have mountains of evidence that show inflammation contributes to autoimmunity as well as many psychiatric disorders.

The discovery of the link between inflammation, autoimmunity, and psychiatric illness has revolutionized our approach to mental health, transforming the field of psychiatry and ushering in the emerging fields of immune-neuropsychiatry and even psychoneuro-endocrinoimmunology (the study of the interactions of the mind, the nervous system, the endocrine system, and the immune system).

Increasingly, doctors in these fields are discovering that they can help ease the psychiatric symptoms of children, teens, and tweens without using psychotropics. They're also learning that many if not most psychiatric medicines contain some anti-inflammatory properties that may in fact contribute to their efficacy. This suggests that some anti-inflammatory pain relievers—for example celecoxib, commonly known under the brand name Celebrex—may be effective in easing the symptoms of neuropsychiatric disorders. Indeed, I have had considerable success using this medication with many of my tween and teen patients who suffer from brain inflammation.

Of course inflammation alone isn't always the root cause of a child's mental health problems, but it can often contribute when the conditions that do underlie their main health issue are inflammatory in nature. For example, hypothyroidism—an underactive thyroid gland—is not always connected to inflammation, as far as we know. But if it's caused by an autoimmune condition known as Hashimoto's thyroiditis, in which inflammation compromises the gland's ability to produce thyroid hormones and which can lead to cognitive and psychiatric symptoms, we'd have to explore a possible connection. It's important to look at the whole picture.

The lower your layers rest in the kettle, the more room your immune system has to maneuver, granting you more flexibility and resiliency. Why does it suddenly seem like our children have become markedly less resilient—developing more allergies, more autoimmune issues, and suffering from more mental health problems? It's not because there's been a dramatic change in our genetics—evolution doesn't move that quickly—but rather because there have been dramatic changes to their environment, which can affect genetic expression. This modification of genetic expression by environmental factors (known as epigenetics) is the key to understanding the effects of the interacting layers in the immune kettle. It's why one person's baseline immune strength might be such that her body and mind can withstand multiple onslaughts of, say, allergies, infections, stress, or poor nutrition and still function well even as the contents of her kettle simmer gently near the top, while another person's equally full pot may boil over even when it's just stress that's added to the mix.

And even then, no two overflows look the same, influenced by factors such as genetics, age, the timing of the exposure, and the duration of the exposure. What manifests as a classic food allergy reaction in one person might present as neurological symptoms like ADHD or cognitive dysfunction with brain fog in another. One child might withdraw and become depressed, while another might begin to ruminate and act out violently. Minor hormonal imbalances can leave one person asymptomatic and another feeling low. Identifying the dysfunction behind a child's symptoms allows us to work on increasing the immune system's resilience—lowering the layers in the kettle before it reaches a boiling point.

Immunity Is in the Balance

Many people have heard that to achieve good health they should "boost" their immune system, a message that might lead them to load up on vitamin C to mitigate the effects of a virus or drink green tea in hopes of staving off cancer. But for many of the kids I meet, a boost to the immune system is the last thing they need. Their immune cells are so active and hypervigilant that they're picking fights indiscriminately, blind to the difference between invaders like harmful bacteria and viruses, ordinary passersby like dust and eggs, or friendly natives like the child's own joints and brain. These kids need a regimen that will calm the immune system, not rev it up. Other kids may have an immune system that is deficient, in which case of course we would want to give it a boost. The ultimate goal, however, is to create a balanced immune system, one that is able to react as needed without overreacting. The principle of integrative medicine is function, not numbers. A lab can tell me a child has a healthy number of white blood cells—your immune system's key fighters against viruses, bacteria, and other attackers—but if those cells aren't functioning properly, their number is irrelevant. Ultimately, a healthy, functioning immune system must be neither too aggressive nor too lax, guarded by well-educated, properly trained, appropriately reactive immune cells that understand the parameters of their jobs. This means my work is often twofold—help identify and eliminate any bad bugs harming my patient, while also bringing the immune system back to a state of balance.

What Balance Looks Like

The immune system is one of the most complex systems in the human body, but at its essence has two jobs: 1) to recognize friend from foe, which can include identifying self from nonself, and 2) to eliminate the foes, allowing for cellular healing. Part of identifying self from nonself is also knowing when to ignore the nonself. For example, food is nonself but requires no reaction. It should be left alone. However, if food is contaminated with harmful bacteria, like pathogenic *E. coli*, the immune system needs to go to work to get rid of the bad actors. The same goes for other types of antigens— substances that provoke an immune response—such as parasites and viruses.

Every millisecond, your immune system is involved in a constant stream of cross talk and feedback between billions of cells, sometimes trying to stimulate a reaction, sometimes trying to inhibit a reaction, all in service of generating the proper response. When the immune system is out of balance, it creates opportunities for miscommunication. And just as in war (or marriage), when communication breaks down, the whole system breaks down. Sometimes the receptors don't work (more on those later), or there's a distraction and something that shouldn't have gotten through breaches a barrier. Sometimes, too, the immune system mistakes a host "self" cell for the enemy and goes on the attack, spraying it with friendly fire. Frequently, the physical manifestation of this miscommunication is inflammation and autoimmunity.

Inflammation in and of itself, of course, is not a bad thing. When we cut ourselves or twist our ankle, the redness, swelling, and pain we experience is a sign that our immune system is properly doing its job, heeding the call of the injured cells by sending additional troops to help clot blood, close up tissue, and eradicate any pathogens that may have snuck in. When we're coming down with a viral or bacterial infection, our skin feels hot as our body responds with systemic inflammation in the form of a fever. Acute inflammation—that is, temporary inflammation that subsides once the battle is done—is healing. Chronic inflammation, on the other hand, is a problem. Whether it's because the immune system's forces just aren't strong enough to vanquish an invading enemy on its own, or a miscommunication means the troops never got the message to pull back once the job was done, chronic inflammation puts a strain on otherwise healthy tissue and organs that can cause a cascade of dysfunction and subsequent symptoms. That lack of communication is often a significant contributor, if not a major contributor, to a child's mood disorder.

We used to believe there was little interaction between the immune system and the central nervous system (CNS), which includes the spinal cord and the brain. The CNS was thought to be "immune privileged," meaning an organ or system that doesn't react to the presence of antigens with an inflammatory response. Today we know that's not true. As we'll discuss shortly, we now know that the barrier separating the brain from the bloodstream (the blood-brain barrier) is neither solid nor impermeable as previously believed. Researchers in the burgeoning field of immunopsychiatry have proven that inflammatory cytokines—messenger molecules released by certain white blood cells when the body is under attack by illness, infection, or stress—can in fact access the brain and cause changes, like inflammation, that trigger mood disorders. Several studies have shown an association between certain inflammatory markers and depression. This means that while a resilient, balanced immune system is fundamental to physical health, it's crucial to our mental health as well.

The Two Branches of the Immune System

The immune system works on two levels: innate or primitive, which you can think of in hugely simplified terms as general and indiscriminate; and adaptive or acquired, which again, hugely simplified, you can think of as learned and specific.

Innate immunity refers to the immune defenses you were born with, the ones that make up that first layer of your kettle. It's nonspecific in that it's going to work to recognize and repel anything it recognizes as foreign or undesirable, no matter what it is. It's primitive; it has no memory and reacts the same way every time it spots something suspicious. Adaptive immunity, which is activated by the innate immune system, offers a more focused and custom response. It's made up of immune cells that are programmed to destroy particular viruses, bacteria, or harmful cells like cancer, some through direct contact, some through the creation of proteins called antibodies. Adaptive immunity has memory. Once these cells eliminate a certain antigen, they'll always recognize that antigen and attack it if they detect it again. That's the premise upon which vaccines work—introduce a small amount of a killed, weakened, or inactive particle of a virus or bacteria, and the immune system creates antibodies to combat it, thus training it to obliterate the pathogen if it ever shows up again.

The way these two branches of the immune system direct their troops—that is, the way the cells in the immune system communicate with each other—is through cytokines, or immune messenger molecules. Like neurotransmitters in the nervous system or hormones in the endocrine system, immune messenger molecules carry information (which is why years ago the neuroscientist and researcher Dr. Candace Pert referred to them as informational molecules). There are many types of cytokines that travel to and from different cells, with multiple targets that often overlap. Some are pro-inflammatory (they provoke inflammation), some are anti-inflammatory (they counter inflammation), and some are regulatory (they help balance inflammatory responses). There are three cytokines you will want to remember: interleukin-6 (IL-6), which is pro-inflammatory; interleukin-10 (IL-10), which is both anti-inflammatory and regulatory; and the powerful inflammatory signaling molecule tumor necrosis factor-alpha (TNF-α), which you might have heard about from ads for biologics (i.e., Humira), prescribed for different autoimmune diseases, because they work by blocking TNF-α. Some immune cells can produce different messenger molecules depending on the type of organism they encounter. The communication lines are so complex, it's amazing that they work as well as they do.

Three Lines of Defense

It's relatively common knowledge that unlike a stand-alone organ such as the liver, the immune system is made up of multiple satellite locations throughout the body, like the thymus, bone marrow, spleen, and lymph nodes (which is why they get swollen when we're sick). What's less commonly known is that it's also present in the skin; the mucosal linings of the nose and throat; and the gut, or gastrointestinal system, which houses as much as 70 to 80 percent of the body's immune cells.

Our bodies are not so much temples as they are medieval castles, with multiple barriers to entry and sophisticated overlaps and strategic redundancies in place to nullify intruders that manage to make their way in. The first line of defense is barrier and chemical protection, and is part of our innate immunity. We're all basically walking around covered in an immunity force field comprised of skin, hair, microbial flora (good bacteria), and complement proteins (which jump onto bacteria and make it easier for immune cells to do their job). The acute inflammation that occurs when something breaches that force field, making us vulnerable to infection—when we cut or burn ourselves, for example—is part of our innate immunity. Protecting our nose, eyes, mouth, and other orifices that would otherwise present easy entryways to the body are nasal mucous (your kids will know it as snot), tears, saliva, and other secretions released by our mucous membranes, all designed to trap and kill pathogens before they can multiply or invade deeper.

In the second line of defense, also part of our innate immunity, we find phagocytes, whose name comes from the Greek word *phago*, meaning "to eat or devour." They have one mission: see the bacteria, eat the bacteria. They're a type of white blood cell, and they're not picky. When they spot something that made it past the first line of defense and could make us sick, they bind themselves to the offending pathogen, engulf it, and destroy it with enzymes. They're soldiers, patrolling our blood and tissue and consuming harmful cells and microorganisms, but they're also the cleanup crew, collecting the old, dead, or dying cells to keep that toxic material from piling up. There are multiple types of phagocytes, but the following three are the ones that provoke symptoms that might be familiar.

Granulocytes get their name from their grainy appearance (caused by granules in their cytoplasm). Neutrophils, the most common, are usually the first on the scene to fight infections. If you see pus coming out of a wound, those are dead neutrophils. Basophils and eosinophils are partially responsible for allergic inflammation that causes symptoms like hay fever, hives, and asthma.

Macrophages are extremely large phagocytes. Their job is to engulf bacteria and eliminate dead or dying cells and cellular debris.

Dendritic cells monitor the epithelium—the single-cell outer lining of the skin and the tissue that lines most of our body cavities, as well as the organs that have contact with outside influences, like the lungs and intestines—for possible pathogens and antigens.

Dendritic cells have little tentacles that can reach out to grab an antigen, draw it across the epithelium, and present it to B cells and T cells—two types of white blood cells that are part of the *adaptive* immune system and are always floating around beneath the epithelium. One develops in the (B)one marrow, the other in the (T)hymus. As they do that, the dendritic cells send one of two signals to the B and T cells: 1) "Foe! Kill it!" or 2) "Friend. Just an innocuous piece of food or bacteria. Let it pass." The dendritic cells essentially guide the T and B cells to mount the proper response to whatever foreign molecules the dendritic cells come across. They help maintain immune balance, or homeostasis.

In addition to phagocytes, natural killer cells (NK cells) also play a major role in the innate immune system. These come by their name honestly—they're cytotoxic, which literally means "cell killing." Activated by cytokines, they scout around in the blood, tissues, and lymph nodes for "physiologically stressed" or damaged cells, such as those infected by viruses or that have mutated into cancer cells. Every cell expresses proteins on their surfaces. But when the receptors on NK cells recognize foreign proteins, they're activated to move in and assassinate the affected cells. They do this by producing pro-inflammatory messenger molecules (cytokines and chemokines) and by latching on to the affected cell and injecting it with cell-killing granules that essentially digest the cell. NK cells are critically important for fighting cancer and controlling viral infections.

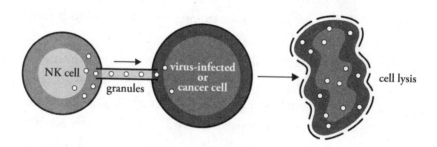

Finally, we get to the third line of defense, which is where our adaptive immunity kicks in. These immune cells can learn—meaning once they eliminate a certain type of cell, they'll remember to attack it if they cross paths with it again. For the purposes of our discussion, we want to concern ourselves with white blood cells called lymphocytes. There are three types: B cells, T cells, and cytotoxic cells.

B cells have receptors that are matched to specific antigens and can only bind to those antigens. After a B cell encounters an antigen, it presents a piece of it on its surface so a T cell can "see" what it is. The T cell then signals back to the B cell with instructions:

1. If the T cell has never seen this antigen before, it will tell the B cell to mature into a plasma cell, which produces the antibodies necessary to destroy the antigen. You'll probably feel sick until enough antibodies are released to handle most of the antigens in your system. Although most plasma cells die once they have eradicated their respective antigens, a subset live longer and provide more long-lasting immunity.

2. Once enough of those antibodies are doing their job, the T cell will start telling other B cells presenting this antigen to mature into memory cells, which are also able to bind to that specific antigen. This is all part of the primary immune response. Further, memory cells don't die after binding to their antigen, they linger. Because they "remember" their specific antigen, the next time it shows up they are able to target and respond to it much faster. Upon this repeat exposure, they will reactivate and start turning into plasma cells that produce very large amounts of high-affinity antibodies, meaning these antibodies are even more sensitive and faster acting than the antibodies produced during the first, or primary, exposure. Memory cells are why you probably won't feel sick during the repeat exposure; they react so fast the antigens don't have enough time to build up in your system. This is what is referred to as the secondary immune response.

Unlike B cells, T cells cannot produce antibodies. Most T cells are involved in regulating the immune response, making sure it's active enough to fight an infection, but not so active that it causes an autoimmune response. When a "naïve" T cell binds with an antigen, it can be activated and turn into any one of a number of different types of T-helper (Th) cells:

- Th1 secretes pro-inflammatory cytokines and is important in cellular immunity related to viruses and fungi, including candida or yeast.
- Th2 is involved in humoral or antibody-mediated immunity. *Humor* is the Latin word for "liquid." Th2 targets antigens that have not yet invaded any cells and are still floating around in the blood and other body fluids, stimulating B cells to make antibodies and to secrete various types of cytokines. An excessive Th2 response can antagonize Th1 and cause the inflammation that manifests as allergic reactions.
- Th17 is the new kid on the block, its discovery as another T-helper cell only dating back to the first decade of this century. It's a heavy hitter, accounting for much of the inflammation we used to attribute solely to Th1. In addition, it is also important in contributing to autoimmunity.
- Tregs: Back when I wrote my first book in 1997, we called these T suppressors. The name changed because whereas it was once thought the job of these T cells was to limit immune function, they are more like harmonic conductors, balancing and regulating the immune system, specifically the production and action of Th1, Th2, and Th17, to help prevent and regulate inflammation and autoimmunity. They are essential for maintaining tolerance to self (an important quality at the cellular as well as whole person level).

Cytotoxic T cells are effector-type T cells, rather than the regulatory-type T cells we've just discussed. They are like another type of cytotoxic cell, NK cells, except instead of killing indiscriminately, these effector cells only target specific antigens.

T-Helper Cell Differentiation

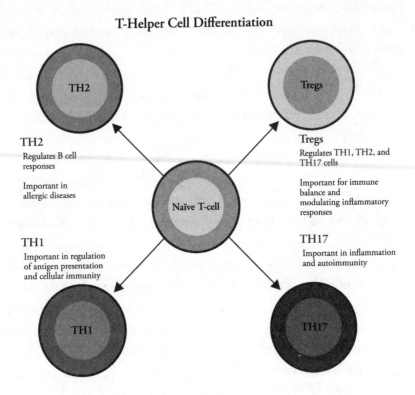

TH2
Regulates B cell
responses

Important in
allergic diseases

TH1
Important in regulation
of antigen presentation
and cellular immunity

Tregs
Regulates TH1, TH2, and
TH17 cells

Important for immune
balance and
modulating inflammatory
responses

TH17
Important in inflammation
and autoimmunity

The processes that take place in the three lines of defense—barrier and chemicals, phagocytes, and lymphocytes—do not occur in linear fashion. Rather, our immune cells work simultaneously in a complex, often overlapping fashion to keep us healthy. In sum, our bodies have developed an excellent defense, repair, and cleanup system supported with lots of backup to keep the immune system well balanced and minimize the chances of any reaction getting out of control.

Yet sometimes the immune system still gets overwhelmed, over-reacts, or loses the ability to differentiate self from nonself and attacks its own organs and tissues.

Autoimmunity

Until now we've focused our attention on the myriad and impressive ways our immune system protects us from harm. Yet we should also be grateful for all the times it has spotted a foreign substance and known *not* to react, a process known as immune tolerance. A fetus growing in a woman's body, for example, is certainly an example of nonself, but we wouldn't want the mother's immune system to reject it. We'd be unable to function if our body reacted to every dust mite that passed under our noses. A healthy immune system knows how to differentiate things that are harmful from things that are not, and to mount an appropriate response. Food, dust, or grass is not inherently harmful—plenty of people don't react at all to them. Why? Again, an example of immune tolerance. Their immune system sees the food, dust, or grass—the antigen—and ignores it. In people who do have an allergic reaction, it's because their immune system decided there was something harmful about that antigen and proceeded to mount a hyperreactive response involving Th2 cells and a type of antibody called IgE.

But sometimes the body violates its primary tenet, loses the ability to differentiate self from nonself, and mounts an immune re-sponse against its own cells and tissues. This is called autoimmunity. Everyone has self-reactive T and B cells, but normally they're kept in check by a healthy immune system. However, when immune cells or antibodies become destructive and attack an organ, that's evidence of immune dysregulation that we call an autoimmune disorder. For ex-ample, when autoimmune inflammation targets the joints, it causes the painful swelling of rheumatoid arthritis. If immune cells attack the insulin-producing cells in the pancreas, it manifests as type 1 diabetes. If they attack the lining of the colon, you get the bloody stools and abdominal pain of colitis. And if immune cells and in-flammatory molecules attack the brain, the resulting inflammation (autoimmune encephalitis) can cause neurological, psychiatric, and mood disorders.

What drives autoimmunity? Frequently, that bottom layer of the kettle, genetics, is the first to blame. Next, anything that creates an imbalance between Th1 and Th2, or activates too much Th17, or decreases Tregs, can contribute to autoimmunity. Our environment is increasingly full of chemicals, toxins, and pollutants that do just that. Poor nutrition can do it, too, as evidence shows that vitamin deficiencies can have a significant impact on the immune system, especially insufficient levels of vitamin A and vitamin D. And as we'll see in upcoming chapters, infections—viral, bacterial, or otherwise—can trigger autoimmune responses, an event that I believe is more common than is typically recognized in the medical community.

The Microbiome

Finally, in recent years it has become evident that the health of our immune system is highly connected to the health of our microbiome—the combined genetic makeup of the trillions of microbes that live on and inside us, from our skin to our gut (we'll explore the microbiome in greater detail in the next chapter). This microbiome rapidly educates our immune system once we leave the safety of our mother's body. In utero, a fetus develops with an immune system that favors anti-inflammatory Th2 responses to antigens, possibly to inhibit the production of the inflammatory Th1, which would normally try to destroy a foreign presence (i.e., the fetus, which is foreign due to half of its genome, and subsequent genome-encoded proteins, coming from its father) in the mother's body. When the baby is ready to be born, its immune system is still heavily skewed toward Th2, but needs to quickly whip up enough Th1 to defend itself against viruses once it's outside the safety of its mother's body. Nor can it survive without Tregs and Th17, which are necessary for healthy, controlled inflammation. As the baby descends through the vaginal canal, it's exposed to the flora and bacteria of the mother's own microbiome, which primes its immune system and kick-starts the development of Th1 and Tregs, which work to start balancing out all the Th2. (My colleague Sid Baker, MD, the king of metaphors, has suggested he'd like to see OBGYNs and midwives purposely introduce some of the mother's vaginal flora into the newborn's mouth, letting them have a "perineal picnic.") One theory for the increase in the number of people with allergies and autoimmune diseases is a rise in the number of C-sections, which eliminates the chance for the neonate to have contact with its mother's microbiome. This can result in a baby with increased Th2 and decreased levels of Th1, Th17, and Tregs and their appropriate inflammatory, anti-inflammatory, and regulatory properties. Babies with excess amounts of Th2 can grow into children, teens, and adults with allergies, eczema, asthma—and an increased propensity to autoimmunity, which as we've seen can be associated with mental health and mood disorders. That's why the earlier we can shore up a child's immune system and microbiome with breastfeeding (if possible), good nutrition, and infant probiotics/prebiotics, the better.

I've known women to break down in tears in my office, consumed with guilt that they did something wrong during their pregnancy or their baby's first few months of life that's to blame for their child's suffering. I promise you, you did nothing wrong, and if you had to have a C-section, you're not to blame for any of your child's mental health issues. You did what had to be done for your safety and the safe delivery of your child. Remember, autoimmunity is only one of a multitude of influences related to the microbiome. It takes a perfect storm of influences and predispositions to bring any disorder into existence, a unique confluence of factors that are often completely beyond our control—like antibiotics, which can lead to dysbiosis (a microbial imbalance) or a yeast overgrowth; environmental exposures; and even sleep quality—to trigger a mood disorder. The potentially increased risks of a C-section should not deter women from choosing the procedure should they and their doctors decide it's necessary. Our interest in rooting out the causes of a mood disorder is not so that we can place blame or debate what we could have done differently, especially since the answer is usually nothing. It's so we can use the information we have to inform treatments and future actions. Throughout this book we'll cover ways to mitigate the effects of any possible contributors to mental health disorders. Since many of these are out of our control, our best option is to focus on keeping the levels in our children's immune kettles as low as possible so their bodies are better equipped to cope with whatever life throws their way.

Inflammation often points to an out-of-balance immune system, and an out-of-balance immune system often points to an out-of-balance microbiome. This implies that in order to learn how to help our bodies—and brains—better cope with the stresses and influences of the outside world, we need to better understand how our insides work. In particular, the gut.

QUESTIONS TO CONSIDER:

Is there a family history of autoimmunity?

Is there a family history of allergies, eczema, hay fever, or asthma?

Did your child's mother receive antibiotics when pregnant, or during labor and/or delivery, such as for Group B strep?

Was your child born vaginally or via C-section?

Did your newborn infant receive antibiotics, e.g., in the NICU during a rule out sepsis protocol?

Was the child breastfed, and did the mother receive antibiotics while breastfeeding?

Has your child exhibited a propensity for allergies?

Did your child get recurrent infections while growing up?

Did your child receive antibiotic treatments frequently through early childhood?

CHAPTER 3

Gut Feelings

For all their bathroom talk and gross-out humor among their peers, in my experience the last thing teenagers and tweens want to talk about with adults, perhaps aside from sex, is their poop and gas. And yet several times per week, this is the topic of conversation I have with my young patients. They'll sit on my couch with their earbuds plugged in, often slouchy and morose, sometimes hyperactive, sometimes spacey and out of it. Regardless, once I start my line of questioning, the expression on their face generally makes it exceedingly clear that they would rather be anywhere but here with me, talking about anything but this.

I'll begin by asking them questions about the frequency and consistency of their bowel movements; they'll avert their eyes and mumble three or four words that I have to strain to hear. I've found the most effective way to break the ice is to lean forward and say, "Hey, you think talking about this is bad? I've had people show up in my office who want to show me pictures of their poop!" It's true. Some families come in waving pictures like excited new grandparents; others just start flipping through their phone and there it is, right next to a selfie. That usually cracks the kids up, and then we start making progress. Because although they desperately don't want to be sitting on the couch talking to me, they also desperately want me to help them get rid of their noisy, malodorous problem.

Being a tween or teenager is hard enough. Being a gassy teen or tweenager is pure hell. Because my office visits tend to last approximately a half hour to an hour and a half, my days are spent consulting in my office or between two exam rooms. More than once, even in the dead of an Upstate New York winter, I've had to throw the windows wide open to clear out the odor some of my poor patients leave in their wake before I can allow a new patient to come in. I've mastered my poker face and would never dream of embarrassing or hurting the feelings of any child who comes to my office with this problem. Most of these kids' peers are not so considerate or kind. So then you have to wonder, is the child's low mood and anxiety a symptom of physical or mental illness, or is he low and anxious because he's so worried about uncontrollable gas and being ridiculed? I'd be low and anxious too if I carried that worry around with me. Wouldn't you?

Those are just the kids with obvious intestinal issues. But you might be surprised to learn that I ask this same round of questions with all my teen and tween patients. Like Alan, an adolescent who seemed plagued with every classic symptom of teenagitis known in the history of teenagers. He was pimply, gassy, sluggish, and morose. Then again, what sixteen-year-old boy isn't? That's not why his parents brought him to see me. Their concern was their son's recurring bronchial infections, despite years of being treated with antibiotics. And even more worrisome, while Alan had often been moody in the past, he was becoming more and more withdrawn. In the words of his mother, a salt-of-the-earth type, he just seemed "out of it."

Like many of the teens in my office, the day we met he looked like he wanted to curl up in a ball and hide. His wavy, light brown hair hung over his eyes. I settled in across from him, leaning over so my elbows were on my knees, and started out as I often do:

"Hey, Alan. How are you?"

"Fine."

This is usually the response I get, and it's not surprising. When most people ask us how we are, we know they don't actually want an honest answer. It's therefore incumbent on me to make it clear to my tweens and teens that I really do want to know, and I really do care. Although it doesn't always pay immediate dividends, it plants a seed that frequently germinates and produces fruit at one point or another.

"Well, tell me, why do you think your mother brought you here?"

"I dunno."

If you have a teenager, you know exactly the tone of voice he used. It's a cross between a grunt and a three-note song phrase. Teens make it without moving their lips, rendering the words unintelligible, and we only know the kid is saying "I dunno" because we used the same tone ourselves when we were their age, as did generations before us.

One more time.

"If I could wave a magic wand and help you with anything that's bothering you, what would it be?"

He slightly shrugged his shoulders. "My mom says I'm acting weird. Not like myself."

We talked a bit about moms and their propensity to overreact and make big deals out of nothing. Then I started asking him about his bowel movements and the chronically smelly gas his mother had told me about. He gave me the same look all the kids give me who think they're seeing me for their inexplicable brain fog, diminished cognitive abilities, hyperactivity, anxiety, or depression. Why in the world would I want to know what's going on inside their butt instead of their brain?

The reason is because what comes out of their butt is indicative of what's going on in their gut, and what goes on in their gut has a huge effect on what's going on in their brain.

The idea that a disorder in one part of the body can manifest as symptoms in another part of the body is not new. While many archival records are so incomplete it's unlikely we'll ever know for sure what caused many of the most famous epidemics in ancient history, we do know that for centuries illnesses, infections, and toxic loads were misdiagnosed as either demonic possession or mental illness. A strep infection can cause a sore throat, but when it initiates a misdirected immune reaction, it also attacks the basal ganglia in the brain, causing people to flail their limbs uncontrollably. This condition, known as St. Vitus' Dance, may, in conjunction with other possible factors such as religious fervor or intense stress due to bad weather and poor harvests, help explain why villages in parts of medieval Europe were occasionally struck by mass "dancing mania," in which citizens contorted and gyrated themselves into exhaustion and even death. Some theorize that toxic levels of ergot, a fungus that could have been prevalent in local grain (and the origin material from which LSD is derived), could have caused the hallucinations and physical torments that drove some young girls in sixteenth-century Massachusetts to accuse their neighbors of witchcraft, especially if mixed with an equally toxic blend of religious pressure and superstition, boredom, guilt, jealousy, or fear of the other. Today, we know that liver disease can cause ammonia buildup in the brain that can look and feel like mania or paranoid schizophrenia, that lupus can cause symptoms of psychosis, and that symptoms of diabetes can mimic depression. And thanks to advances in technology, we've also confirmed that our gut plays an enormous role in our general and mental health.

Practitioners in my field of integrative medicine have known about the importance of gut health for over a century. In 1904, Élie Metchnikoff, who ran a lab at the Pasteur Institute, triggered a run on yogurt in France after suggesting it could be used to increase the amount of good bacteria in the gut, which his research indicated played a large role in human health and longevity. (Four years later he was awarded the Nobel Prize in Physiology or Medicine for work conducted long before he took an interest in yogurt—his discovery of phagocytes, the white blood cells we discussed in the last chapter that "swallow" bacteria and clear the blood of dead or dying cells.)

The notion that the flora in the gut could affect overall health was questioned and debated for years because the connection was impossible to measure. Not anymore. In the last five years or so, the extraordinarily precise metabolic readouts made possible by mass spectrometry—far more comprehensive than anything we could glean via cultures grown in a petri dish—allow researchers to paint a vastly more detailed picture of the ecosystem of microbes (aka the microbiota) and their DNA in the gut, collectively referred to as the microbiome, than they could before. For example, it was once thought that the ratio of microorganisms in the gut to human cells was something like 100 trillion to 10 trillion (10:1). Now we know that ratio may vary, and at times be closer to 1:1. This means that it's possible that with every "defecation event," to put it politely, you decrease the number of microorganisms in your gut to the point you've almost flipped the ratio so that human cells outnumber microorganisms. That's right, pooping can be transformative! It also explains why constipation and diarrhea take such a toll on the body—they throw your body's microbiome out of balance.

Thanks to metabolomics—the study of the end products and byproducts of the metabolic processes that break down our nutrients and turns them into energy—we can monitor the effects of microorganisms in the body by tracking their metabolites, small molecules produced by the microbiota that are involved in the development and modulation of the host's physiology, specifically its immune system. For instance, when we look in the stool or urine of a patient suffering from psychiatric symptoms and find certain metabolites present that can affect the brain, it can be a clue that though the brain is displaying the symptoms, the organ itself is not the birthplace of the mental or behavioral disorder.

Being able to so closely trace the movement of metabolites throughout the body has also confirmed the existence of bidirectional connections between our immune system, our gut, and our brain— sometimes referred to as the gut-brain-immune axis. It's astounding how much information this intricately intertwined and shockingly complex triad is processing every microsecond. When things go wrong in the microbiome, that interwoven complexity can contribute to dysfunction such as inflammation and autoimmunity, and can be seen in other organs besides the gut. Acknowledging this connection—including the gut's influence on the lungs, joints, liver, bones, and many other organs and systems—is essential to our ability to treat disease and dysfunction. An assault on one is a possible assault on all. It's therefore not enough to target our efforts on simply eradicating the agents of stress, which can be both psychological and environmental. We also have to strengthen the host, specifically where those microbes reside. This means that sometimes in order to heal our children's minds, we need to go straight for the gut.

The Gut

When most of us think of the gut, we think about the stomach and intestinal area, where we typically imagine digestion taking place. It's also where we tend to associate that unnerving "gut feeling" when something isn't quite right, or we're worried, scared, or stressed (there's a reason you experience this feeling in your stomach and intestines, though, and we'll get to that in a minute). But in reality, the term *gut* refers to the entire gastrointestinal tract and runs from our top to our bottom, encompassing the mouth, esophagus, stomach, small intestine, and large intestine, which includes the colon and rectum.

Along with your skin and respiratory tract, the GI tract stands as one of the body's three major points of interaction with the outside world. It's lined with approximately 100 trillion bacteria—about four pounds' worth—with the majority residing in the large intestine. This is the microbiome. The microbes in the gut are constantly barring, tolerating, and battling the billions of foreign microorganisms with which our bodies come into daily contact. In fact, the microbiome is so vast and its function so influential, some scientists say it could be considered a separate "ecosystem organ."

On its own, the GI tract consists of more surface area than your skin and respiratory system combined. While it includes the mouth, esophagus, and stomach, our discussion will center on the intestines because that's where most of the action takes place. In fact, if you were to flatten and smooth out the twenty-six feet of intestinal tract coiled up in our abdomen, the surface area would approximate that of a regulation singles tennis court. That's twenty-six feet of intestines providing millions of nonstop opportunities for absorption and immune interaction.

The GI tract serves multiple purposes and has three basic digestive functions. First, the gut wall acts as a barrier, keeping harmful substances from escaping into your bloodstream. Second, the gut is where digestion takes place and food is broken down into components that can be used by the body. And third, the gut is where nutrients from food are extracted and absorbed.

The Gut-Immune Connection

To increase the surface area and allow for more absorption of nutrients, the lining of our intestines is hilly, like a mountain range. The high points are called villi and the low points are called crypts. And all along that lining is the intestinal epithelium, which, as we discussed in the last chapter, is the single-cell lining that works simultaneously to absorb nutrients the body wants and act as a barrier to the microorganisms it doesn't. The epithelium secretes immunoglobulin A (secretory IgA), an immune molecule whose job is to recognize invaders and intercept them even before they can attach to the lining. If the foreign material does make it to the lining, the dendritic cells (which we also talked about in chapter 2; if you skipped it, I promise it's not as dry as you think it will be, and it will be helpful as we move forward) are there, ready and waiting. Imagine the cells along the epithelium lined up like a row of perfect teeth with minuscule spaces in between that only microscopic dental floss could get through.

Antigens can get through the gut lining in one of two ways. As microorganisms approach, certain epithelial cells, M cells, will fold around antigens and pull them through themselves (transcellularly) to waiting antigen-presenting cells, which then present the antigen to T and B cells. The T cells and B cells then confirm whether the antigen is a friend or foe.

The other way antigens can get through is paracellularly, that is, between the cells. Dendritic cells will reach out into that miniscule space between the cells that make up the epithelial wall—which are lined up closely together in what are known as tight junctions—and pull an antigen through to present it to the T cells and B cells in the tissue below. These tight junctions function like turnstiles or gates, slowing the crush of incoming microorganisms enough to allow for healthy, regulated absorption. Without that regulation, everything could come in at once, making the environment impossible to monitor and control, and disrupting the delicate balance of the microbiome.

Sometimes those tight junctions do get loosened, though. Factors such as stress, harmful bacteria, viral infections, and certain medications, drugs, chemicals, and foods can produce inflammatory cytokines (immune messenger molecules) that, along with histamines and hormones, have the power to loosen the tight junction between the epithelial cells. This causes abnormal intestinal permeability, a condition colloquially known as *leaky gut*. As these openings in the gut wall become wider, antigens that normally wouldn't be able to breech the wall can make it out and into the rest of the body. The T and B cells, identifying these intruders as foreign and harmful, mount an immune response. If the cytokines continue to build and the T and B cells become overwhelmed by the number of inflammatory microorganisms, metabolites, and messages pouring in, they can overproduce pro-inflammatory immune messenger molecules, which can lead to inflammation and even autoimmunity in organs beyond the gut, including the brain. This can be fueled by an imbalanced microbiome spiraling out of control, a condition called dysbiosis.

Dysbiosis isn't caused only by the most threatening pathogens, like salmonella. Inadequate nutrition or an overgrowth of anything like yeast or other less pathogenic bacteria could induce gut dysfunction or disease. This dysbiosis can alter the immune system, changing the way it responds to pathogens. The change can happen slowly, little tweaks here and there that eventually snowball enough to alter into something bigger or simply shift the relationship between the microbiome and the rest of the body. Once that relationship is out of balance, other systems and organs—including the brain—can start to suffer.

The Gut-Brain Connection

In addition to the 70 to 80 percent of your immune cells and the approximately four pounds of bacteria found in the gut itself, inside the lining of the gut stretches a web of about 500 million neurons that makes up what's called the enteric nervous system, which is so active and handles so much information, many scientists refer to it as our "second brain." The enteric nervous system is responsible for enabling and managing the contractions, absorptions, and secretions necessary for proper digestion. It also maintains a bi-directional relationship with the central nervous system, linking "emotional and cognitive centers of the brain with peripheral intestinal functions" by sending messages back and forth across the vagus nerve, a major player in the parasympathetic part of our autonomic (involuntary) nervous system, and the longest cranial nerve, running all the way from the brainstem to the abdomen. This explains why we often physically respond to emotional stress through our gut—we literally feel it there! If you've ever had to run to the bathroom before taking a test, or felt butterflies in your stomach before a performance or presentation, or vomited after hearing distressing news, you can thank this relationship between your gut and your brain.

The brain communicates messages that alter the production of intestinal motility, intestinal permeability, immune function, mucus, and biofilm (a "community" of bacteria that can coat surfaces— including the intestinal lining—with a slimy protective layer resistant to antimicrobial and antibiotic treatments), which consequently changes the composition of the gut microbiome. The communication works in reverse, too; what happens in the gut affects the brain. As we're about to see, there is ample evidence of a connection between the microbiome and the brain.

The microbiome controls the development and maturation of the microglia, the brain's resident macrophages, which are, as you'll recall, part of the innate immune system. Gut microbes also produce neurotransmitters, like GABA, an inhibitory neurotransmitter that can ease anxiety, and serotonin, also known as one of the "happy hormones" —in fact, the gut produces about *90 percent* of the body's serotonin! And as we've discussed, the microbiome preserves the integrity of the gut wall with tight junctions.

The thing is, tight junctions don't only exist in the intestinal lining. They are also present in the blood-brain barrier.

The Blood-Brain Barrier

The blood-brain barrier (BBB) is a multicellular network that acts as a protective intermediary between the blood and the central nervous system, controlling the flow of nutrients and immune cells from the bloodstream to the brain. Like the intestinal epithelium, the BBB provides a selective physical and biochemical blockade that allows nutrients in and keeps waste and toxins out, thus protecting the brain and nervous system. The key word here is "selective." The BBB is not the equivalent of a cement wall, but rather a dynamic, living barrier.

Earlier we compared the body to a medieval castle. The brain is the king's quarters, with the BBB acting as an even stricter guard over what goes in and out than the gut lining. By the time the blood gets to the BBB, the other units of the body's defense should have spotted and eliminated or neutralized any foreign or harmful antigens, inflammatory cells, drugs, hormones, and infectious agents, leaving nothing but healthy, absorbable, brain-friendly molecules of oxygen, glucose, and other essential nutrients to get in. Of course, as we've discussed, sometimes antigens, toxins, or inflammatory molecules do make their way into the circulation and sometimes as far as the BBB. But even when this happens, the BBB is usually able to prevent these harmful molecules and cells from making their way in.

Unless something loosens the blood-brain barrier's tight junctions.

Leaky Gut, Leaky Brain

A disruption in the gut, triggered by, say, repeated exposure to certain foods or overuse of antibiotics with subsequent dysbiosis or candida (yeast) overgrowth, can stimulate the immune cells and cause them to produce large amounts of pro-inflammatory cytokines. This in turn can contribute to dysbiosis and loosening of the tight junctions in the gut's lining. If the gut wall becomes permeable enough, it can allow large inflammatory molecules to spill out into the bloodstream, where they can then travel to the brain. Once there, they can start hammering away at the blood-brain barrier (BBB), loosening the tight junctions there as well.

This concept, colloquially referred to as *leaky gut, leaky brain*, is new and still somewhat controversial in the medical community, but more and more research supports this connection. Not that long ago, the idea of intestinal hyperpermeability was hotly debated, but in the last few years it seems to have gained widespread medical acceptance. I suspect the same will eventually happen with the leaky gut, leaky brain concept. That said, there are plenty of disrupters to the blood-brain barrier that are already commonly acknowledged and accepted, such as infection, oxidative stress, inflammation, radiation, brain injury, and poor diets that are low in protein and high in refined carbohydrates and saturated fats. If dysbiosis and its subsequent inflammatory mediators cause the BBB to become more porous, eventually the inflammatory molecules, including immunoglobulins, cytokines, and immune cells that make their way there, can now break through. Recently, the discovery of immune cells in the meninges (a covering in the brain) and a network of lymphatic vessels called the glymphatic system offered new evidence that the brain isn't immune privileged. The job of this brain "clearance system" appears to be to remove potentially neurotoxic waste products that accumulate in the central nervous system when we're awake. They are essential to getting a restorative night's sleep, which is when they clean and refresh the brain. (The American Academy of Sleep Medicine asserts that tweens should sleep 9 to 12 hours per day, and researchers claim teenagers need exactly 9.25 hours to function optimally. So create a plan with your kids to get the electronics out of their rooms at night and aim for regular bedtimes as frequently as possible.)

What we learned is that large particles can make it through the BBB, and when they do, the information is shared with the immune system in the rest of the body. If the glymphatics can't clear out the large inflammatory molecules from the brain the way they're supposed to, however, those particles can start to activate the microglia—the brain's immune cells—which then go on the attack, releasing their own inflammatory messenger molecules. This attack can result in neuroinflammation and ultimately in the degradation of otherwise healthy brain cells. Now an imbalance that generated an inflammatory response in the gut or the immune system has inflamed the brain. In time, depending on a person's genetic predispositions, nutritional deficiencies, or metabolic or hormonal imbalances, the immune kettle may boil over, at which point we can start to see the behavioral dysregulation on the outside reflecting the dysregulation being experienced on the inside. It can look like depression or anxiety. It can look like a mood disorder.

Evidence abounds that where there is dysbiosis, there are often cognitive and emotional symptoms as well. In the lab, mice whose microbiomes have been altered to produce dysbiosis exhibit anxiety-related behaviors. Many studies have shown associations between dysbiosis and autoimmune disorders and psychiatric disorders, including depression. Some of the more common psychological side effects of antibiotics, which are known to alter the microbiome, include depression, anxiety, and panic. A classic medical school example of a brain disorder due to gastrointestinal dysfunction is hepatic encephalopathy, a consequence of a diseased liver that can't remove toxins from the blood. Symptoms can include mental confusion and lethargy, and in advanced stages, even stupor and coma. Treatment includes specific laxatives, which clear the gut and in turn ease the cognitive dysfunction. Brain fog—a constellation of similar but milder symptoms—is frequently seen in patients with fungal dysbiosis, frequently referred to as a yeast overgrowth, or a chronic candida problem. When we treat the fungal dysbiosis patient with an antifungal and a probiotic, and prescribe a yeast-free diet that removes the sugar and yeast from the system and rebalances the microbiome, gradually the brain fog will clear. As we can see, the existence of a gut-brain connection means that we must be aware that we can't exclusively target the brain to fix what looks like a brain illness, but must instead consider addressing possible gut dysfunction as well.

The Gut-Brain-Immune Axis

We've seen how the gut and immune system work together. We've seen how the gut and the brain regulate and affect each other. We know that the immune system constantly regulates the brain and CNS, and that the brain and CNS can drive immunity. This gut-brain-immune axis represents a vast intersection between immune cells, nerve cells, and cells in the gut, with the microbiome central to them all, as depicted below.

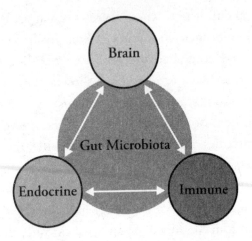

A Rebalancing Act

These three key systems are inextricably interconnected; when one is out of balance, the ramifications can be felt far and wide across the body—sometimes with lasting impact. Which brings us back to Alan.

I ran a few tests that revealed a yeast infection (fungal dysbiosis) in his gut; I suspected this was triggered by the repeated cycles of antibiotics Alan had taken for his chronic bronchial infections. Treatment would include a yeast-free diet, which includes avoiding sugars and refined carbs, as well as a prescription for an antifungal medication and high doses of probiotics. I knew that like many teenagers, he was likely going to resist changing his diet.

Over the course of several visits, I planted the seeds of trust. Beneath the defensiveness and orneriness, I could see hints of the sweetness his mom had assured me used to be part of his disposition, even when he wasn't feeling all that hot from the bronchial infections. So once again I sat down with him. At first we talked basketball. I told him my son was a point guard in high school. We discussed the intricacies of the point guard position, and he told me about some of his NBA idols. I love these types of conversations with kids, but it also helps loosen them up, allowing for a little crack in their defenses.

Alan eventually started to relax and even allowed a tiny smile. That's when I talked to him straight, telling him I couldn't give him a magic pill to fix his issues overnight, but if he was motivated to get better and stick to his treatment even when his mom or dad weren't around to enforce it, it could make a huge difference in his quality of life. I asked him to imagine no more bloating, discomfort, and gas. No more acne. The ability to concentrate and pay attention. Less sadness and anger. I ended with, "If you're not going to commit to this and follow through, though, please don't waste your parents' time and money, or my time."

I knew I wasn't going to get a wide smile and a "Yes, let's go for it!" That would have been too much to hope for. But he did say he'd try.

When I saw the family again, his parents noted that Alan's recurrent bronchial infections had stopped. They also remarked that his concentration and brain fog had improved, and he was doing better in school. That's because when we rebalanced Alan's gut through diet, the antifungal medication, and the probiotic, we began to reverse the negative spiral incited by the yeast overgrowth. When we decreased the inflammation, we helped restore the integrity of the tight junctions of the gut lining, and ultimately that of the BBB as well. And that translated into decreased neuroinflammation, bringing Alan relief from his brain fog, poor concentration, moodiness, and depression. We had extinguished the flames successfully and brought his immune kettle down to a gentle simmer.

In addition to his parents and teachers noting the positive changes, Alan also admitted he was feeling better. But the part he was happiest about was that his belly had flattened out because he was no longer full of gas. Admittedly, no longer being afraid to embarrass yourself in front of your peers would perk up anyone's mood, which will serve as a reminder that any time you're treating psychological symptoms, especially when dealing with teenagers, it's vitally important to take social influences into consideration in addition to noting biological clues. Still, when taken in conjunction with Alan's improved cognition, energy, concentration, and mood, it was an exciting and dramatic turnaround that without a doubt started when we reset the balance of the microbiome and gave his gut-brain axis a chance to restore order and communicate properly again.

Could Your Child Have Gut Dysbiosis?

Of all the questions you'll have to answer in this book, these might be the most uncomfortable. It's probably been a while since you followed your child into the bathroom, so to uncover clues about how well her gut is working might require some awkward conversations. There may be no question a teen or tween is less willing to answer than "Are your bowel movements malodorous?" "How are they shaped?" "Are they firm, soft, or loose?" "How often do you have one?"

OTHER QUESTIONS TO CONSIDER:

Does your child complain that their belly feels bloated?

Does your child crave sugar and carbohydrates, such as junk food, french fries, grilled cheese, sweet cereals, cookies, or potato chips?

Does your child eat a lot of sweet fruits, such as grapes, cherries, and watermelon, or dried fruits, such as raisins?

Does your child drink a lot of soda, fruit juice, fruit juice drinks, or Gatorade or its equivalent?

CHAPTER 4

Out of Control
Infection-Triggered Autoimmune Brain Inflammation (ITABI)

It was happening again, and this time Natalie Maxwell wasn't just feeling frightened and frustrated. She was furious. She'd been in her bedroom folding laundry when she'd heard the now-familiar keening downstairs. She froze, clutching a pair of jeans to her chest, willing the sound to stop. Natalie knew her husband, Brett, had gone down just a minute ago. Surely he would step in. She waited, hoping to hear his low, soothing baritone, which would mean he was working on controlling the problem. Instead, the sound morphed, moving from a high-pitched cry to a full-throated roar. Natalie dropped the jeans onto the bed and walked down the hall. Her ten-year-old daughter, Georgia, stood at the bottom of the stairs, her eyes squeezed tight below her dark eyebrows, mouth open in a howl as tears rolled down her face. "What's going on?" Natalie asked, but it was more out of habit because she knew there'd be no answer.

As Georgia started up the stairs almost robotically, her arms rigid at her sides, her hands clenching and unclenching, Natalie spotted her son, Danny, dart from behind his sister to disappear from view.

And suddenly Georgia was on top of her, flailing her arms and kicking at the walls. To keep the small girl from falling back down the stairs or hurting herself or the house, Natalie grabbed Georgia in a bear hug and wrestled her to the ground, hoping the pressure and forced stillness would contain some of her daughter's inexplicable rage. As Georgia continued to kick and writhe like a fish out of water in her mother's arms, Natalie's thoughts went vengefully to her husband, whom she was 100 percent sure could hear what was going on. Where the hell was he?

In a downstairs hallway, Brett held his son. He'd been on his way to see what was happening to his daughter when Danny had come careening down the hall, colliding with his dad and blocking his way. Upon contact, Danny wrapped his arms tightly around his father's waist and buried his head into his belly. "Dad, when is this gonna stop?" the boy sobbed, his words muffled by his father's shirt. "How can we fix Georgia?" Brett bent over his boy's head, working too hard to keep down his own tears to reply.

The Maxwells could tell you the exact moment their daughter fell ill as surely as if they'd seen her abducted from their suburban street via hidden security camera. And it really was like an abduction. On March 18, 2015, Natalie and Georgia had cuddled for a few minutes before kissing each other goodnight in a flurry of I-love-yous and I-love-you-mores. Natalie had turned out the light and left her daughter tucked cozily in her bed, surrounded by butterflies on the walls and an impressive row of trophies on the shelves, mementos from the various sports at which Georgia, a star athlete, consistently excelled. Natalie went to bed herself not long after. Her job as a consultant for a medical supplies company would have her on the road again in the morning, and she wanted to get as much rest as she could.

The next day, March 19, Natalie went back upstairs to wake Georgia for school. The girl was curled in on herself like a hedgehog, the covers trapped between her hands and knees. Natalie shook the sleeping girl's shoulder.

"Hey, honey, time for school."

Georgia opened her eyes and looked at her mother with an expression Natalie didn't recognize, a dark, hollow stare.

"Are you okay?" Natalie chuckled, thinking Georgia must still be half asleep.

"I'm fine," Georgia replied softly.

Then her voice rose. "Mommy, Mommy, I can't go to school. I can't. I can't go to school! I can't!"

Natalie, who'd been about to leave, turned to face her daughter. This was unusual. Even with pneumonia and a 103-degree fever, Georgia was the kid who'd insisted she was well enough to go anywhere.

"What's going on?" Natalie asked.

Now Georgia was starting to get agitated. Her eyes, strangely dilated and thus darker than usual, darted frantically across the room as she worked herself into a frenzy. Her words tumbled out, erratic and incoherent.

"I can't leave you. I can't go to school. I can't go to school. I can't breathe!"

Natalie sat down on the bed.

"Okay, take a deep breath and calm down. Did you have a nightmare?"

Georgia's expression shifted from terror to anger. "No! I can't breathe! You have to take me to the emergency room."

Natalie could see that Georgia was breathing just fine. There was clearly no obstruction. "Honey, I think you're having a panic attack. Do you have a test in school you're supposed to take that you're nervous about?"

"No."

"Are you afraid of someone? Is someone bullying you?"

"Noooo!" Georgia yelled, her hands scrabbling at the bedsheets.

The clock was ticking, and they were both going to be late. Natalie spoke as calmly and reassuringly as she could.

"Tell you what. Why don't you get dressed, come downstairs, and have breakfast? Maybe you'll feel better."

"No, no, no! I can't eat! Something will get caught in my throat."

Natalie's heart sank. This again?

Two weeks earlier, at a hockey game, Natalie, sitting with Danny, had gotten a phone call from Brett, who was sitting in a different section of the arena with Georgia. Apparently Georgia had choked on some popcorn and wanted her mother.

"Is she okay?" Natalie had asked.

"Yeah, she's fine. She didn't really choke. I mean, she said she was choking, but she could talk, you know? She swallowed the popcorn. She's just really shaky and wants you."

"Maybe her blood sugar dropped? Get her a soda and then bring her to me. We can swap kids."

Brett had brought Georgia over and left with Danny. Georgia had calmed down, but Natalie noticed she was still shaking all over and clearing her throat every few seconds.

In the days that followed, the throat clearing continued. Natalie looked inside her daughter's throat and thought it seemed a little red, but there didn't seem to be any real irritation, and Georgia said it didn't hurt. Still, Georgia was refusing to eat, putting her fork down at every meal, insisting that any food that went into her mouth made her choke. Natalie called Georgia's pediatrician and scheduled an appointment.

The doctor did a throat swab for strep. The results came back negative, but since false negatives are common, and Georgia's symptoms were so odd, the doctor sent a throat culture out to a lab. Bingo! Positive for strep. Natalie and Brett were relieved. All she needed was a few days of antibiotics, they figured, and Georgia's odd throat problem would be gone.

The pediatrician prescribed a seven-day course of Keflex, an antibiotic, which Georgia finished on March 16. Now it was the nineteenth, and she'd woken up a different person.

Natalie was growing frustrated. She informed Georgia it was time to catch the school bus and for her to leave for her business trip. At that, any restraint Georgia had been showing fell apart. She was panicked, feral. Her screams and sobs hurt Natalie's ears.

"You can't leave! If you leave, you're gonna die! You're gonna DIE!"

Georgia managed to get on the bus that day, and Natalie took her business trip, but from then on, not a day went by that Georgia's separation anxiety didn't spike. Whenever Natalie would try to leave, or Georgia was told she had to go to school, the child would fall to the floor, either taut as a wire or curled into a fetal position, clenching her fists in rage. Sometimes the only sound she'd make was the grinding of her teeth; other times she'd scream and cry wordlessly. But often when she'd collapse unexpectedly, it was to shriek in agony. "Something's wrong with me. Help me, Mommy! Something is wrong with my brain! My brain, my brain!" Natalie, knowing no one would believe how intense these episodes could be, started recording them with her phone.

Every other day or so Natalie could get Georgia to school by midday. Georgia's teacher would meet them outside the building and escort Georgia inside, cuddling her and whispering reassurances that no one was going to die or get hurt. Most of the time, though, Georgia stayed home. She developed intensely detailed rituals to keep the worst from happening. If she didn't clear her throat, Natalie would die. If she didn't roll her eyes a certain way, Natalie would die. If she didn't flail her hands, Natalie would die. And if she had three carrots with her lunch, she herself would die. The number three became like poison.

One of the most disturbing compulsions began early. The family had managed to get Georgia to her softball team's opening day. Georgia had been acting strangely for about a week, but as Natalie watched her daughter laughing and talking with her friends, she thought maybe the worst was over. It was a beautiful day, and everything seemed to be fine. Next thing she knew, though, Georgia was standing at her side. Bending her head toward her mother, she whispered, "Mommy, am I in Heaven or in a dream? This doesn't feel real."

Natalie tried not to show her surprise on her face. "You're on Earth," she replied.

"Prove it," said Georgia.

Easier said than done, it turned out. From then on, Georgia asked that question—*Am I in Heaven or in a dream?*—close to twenty times a day, like a mantra, convinced that if she didn't ask it, she would die. She also constantly questioned whether Natalie was really her mother. Natalie had to come up with a code word—lollipop—to prove to Georgia that she was, in fact, her mother, and that she was, in fact, real. If Natalie could give her the password early enough as Georgia started to cycle, she could sometimes keep her daughter from crossing over the line from mere anxiety to full-blown rage. Natalie tried to keep track of potential triggers, but there seemed to be no pattern to her daughter's mood swings. She'd be laughing one minute and screaming the next. No one could figure out why.

For Natalie, the only thing worse than facing the daily onslaught of Georgia's bizarre behavior was facing her family's skepticism. Natalie's in-laws had hinted that there was no problem that a little more discipline wouldn't solve. Even Brett, her husband, was convinced Georgia was putting on a show. The day of the softball game, Georgia became completely unglued after returning home. Brett took that to mean that Georgia could control her erratic behavior. Clearly, as long as Georgia was doing something she wanted to do, like hanging out with her friends, she could keep it together. She was only acting out when she was bored at home and wanted attention. Natalie insisted he was wrong; it wasn't like that. He wasn't home with Georgia all day. There was no way a ten-year-old girl would willingly put herself through so many ritual behaviors and frenzied fits that gained her nothing and left her drained and miserable.

Brett changed his tune the day he told Georgia, who'd been sitting quiet and still on the trampoline in the backyard, that he'd seen her neighborhood friend in the street and invited her over to play in a few minutes. To his surprise, Georgia's response was to cry out, "I can't have anyone here!" and hurl herself off the trampoline, lock herself in her room, and scream for two hours.

Natalie knew her in-laws finally understood the extent of her concern the day she got a hysterical phone call from her mother-in-law. Not wanting to trigger new anxiety, Natalie had snuck away on a business trip without telling Georgia, leaving the girl at home in her grandparents' care. When her grandmother informed her that Natalie was gone, Georgia had started punching the dining-room windows. Now she was locked in the bathroom. "We don't know what to do," said Natalie's mother-in-law. Natalie had her mother-in-law slide the phone under the bathroom door, and talked Georgia out. When she finally had her mother-in-law back on the line, she couldn't help but ask, "Now do you believe me?"

Within a few short weeks, Georgia went from refusing to go to school to refusing to leave her room. She'd spend hours there making sure the comforter was arranged just so, or that a little angel figurine, a gift from her recently deceased maternal grandmother, was positioned in exactly the right spot. Even if she had made it to school, she couldn't have done the work expected of her. The girl who loved to draw could no longer control a pencil or crayon to color, much less properly form letters. Her handwriting had become illegible.

After repeated visits, Natalie's pediatrician refused to do another throat culture, insisting that Georgia wasn't sick, but suffering from normal childhood anxiety, likely caused by Natalie herself. Natalie was flabbergasted. There was nothing normal about what was going on. Her child was wasting away for lack of food and sleep, a shell of who she'd been a few weeks earlier. Half the day Georgia would just sit and stare into space. Her daughter was losing her mind, and Natalie thought she was close to losing hers, too.

A random, lucky text exchange put Natalie in touch with her first reason for hope. It was Easter, and as they did every year, the family traveled to Natalie's hometown of Chicago. Normally Natalie would go out with her girlfriends to celebrate one of their birthdays around that time, but this year, she texted the birthday girl and made her apologies. She couldn't leave Georgia, or rather Georgia wouldn't let her leave her side. She added a few lines about how worried she was and that her pediatrician thought she was overreacting for believing this was something other than garden-variety anxiety. Normally she wasn't in the habit of talking about her kids' health issues, but it so happened that a few weeks ago she had told this same friend, a pediatric nurse, that Georgia had come down with strep.

The text that came up on her phone changed everything.

"Have you ever heard of PANDAS?" Accompanying it was a link.

Natalie had no idea what her friend was talking about, but she opened the link. And there it was. The symptoms coming out of nowhere, the mental confusion and disconnection (known as disassociation), massively debilitating obsessive-compulsive disorder (OCD), separation anxiety, fear of death, fits of rage—everything she'd witnessed in her daughter over the past three weeks was included in the description of this strange syndrome her friend had sent her. She learned PANDAS is frequently referred to as a rare autoimmune syndrome, although it's difficult for me to think of it in these terms as I see so much of it (or its related conditions) in my practice. Some researchers even suspect it may make up 10 or more percent of childhood-onset OCD and tic disorders.

Autoimmune diseases are among the most common illnesses in the United States. On the conservative end, the National Institutes of Health (NIH) believes they affect around 23.5 million people; the American Autoimmune Related Diseases Association (AARDA) estimates it could be as high as 50 million. Around the world, rates of type 1 diabetes, multiple sclerosis, and celiac disease are on the rise in ways that preclude a simple increase in better diagnosis.

The next day, Natalie caught a flight to Florida for her company's regional sales meeting. She was so distracted thinking about PANDAS that the doctor leading one of her trainings, Jane O'Connor, asked her if everything was all right.

"Do you know anything about PANDAS?" she asked.

"Very little," replied Dr. O'Connor. "But I have a friend who works with a lot of PANDAS kids."

Dr. O'Connor got in touch with her colleague and reported back to Natalie that she should get Georgia's blood drawn and conduct an antistreptolysin O (ASO) titer (the concentration of an antibody in the blood) test. Natalie left the conference, sending an apologetic email to her boss and colleagues explaining why she had to get back to her daughter. Her mind wasn't on her work anyway.

Knowing that her own pediatrician would never order the test, Natalie called one of the doctors she worked with back home and begged him to write a prescription for Georgia to get her blood drawn. By the time she landed, her husband had picked up the script, and they were able to get Georgia to the local lab for blood work. A few days later, Natalie got the results. She called Jane O'Connor.

"It says the levels are above three hundred. What does that mean?"

Dr. O'Connor's voice was warm as sunshine.

"It means you've figured it out."

Mystery Illness

In 1998, a clinical research team led by Susan Swedo, who was then chief of the Pediatrics and Developmental Neuroscience Branch of the National Institute of Mental Health (NIMH), published a paper in the *American Journal of Psychiatry* describing a new syndrome named PANDAS—pediatric autoimmune neuropsychiatric disorders associated with streptococcal infections. The study grew from work Dr. Swedo had done earlier in her career on a rare neurological disorder called Sydenham chorea (SC). Her colleague, Dr. Judith Rapoport, chief of the Child Psychiatry Branch at NIMH (and the author of the bestselling popular book on OCD, *The Boy Who Couldn't Stop Washing*), had found literature noting that children suffering from Sydenham chorea were showing OCD symptoms two to four weeks before the involuntary, jerky movements of the arms, legs, and face that comprise the illness kicked in. Sydenham chorea is caused by the same Group A streptococcal bacterium that causes rheumatic fever, an inflammatory disease that, left untreated, can damage the heart. But Swedo and her colleagues found that children with rheumatic heart disease did not show signs of OCD, tics, separation anxiety, rapid cycling mood swings, and aggression, whereas some kids with Sydenham chorea did. Why?

In addition to a study devoted to answering that question, they also launched a treatment study to see what kind of procedures and medicines might help ease the symptoms of the kids with chorea. A doctor sent Swedo's team one of his chorea patients to participate, but on closer investigation, they discovered that the boy didn't have chorea. His movements weren't random; they were exactly the same every time. He wasn't jerking uncontrollably, he was practicing a ritual to throw off the germs he thought he was carrying on his body. And his mother, a medical technologist, had noticed that whenever the boy's older brother, who had Tourette's, came down with strep, *his* tics would get worse until he was treated with antibiotics, at which point they would subside. Somehow, the strep bacteria seemed to be triggering OCD symptoms.

A few months later, Dr. Swedo was studying a group of children with OCD and discovered that a quarter of them had seen their symptoms develop abruptly, perhaps over the course of one to two days, and they were often preceded by a bacterial or viral infection. Since the researchers had noticed that 60 to 75 percent of their Sydenham chorea patients showed symptoms of OCD before they showed symptoms of SC, they decided to concentrate their observations on the children who had developed OCD symptoms following a strep infection. They found that while most SC patients showed signs of OCD, none of the patients with sudden-onset OCD had SC. Dozens, however, were shown to have had strep.

The vast majority of kids get strep throat at some point in childhood and get better, usually with treatment, but sometimes even without. Symptoms are not just a sore throat, but usually a combination of sore, red throats, fever, and swollen lymph glands under the jaw or on the upper neck (and sometimes even atypical symptoms such as headache, vomiting, or abdominal pain). In a subset of kids with a genetic and perhaps environmental vulnerability, the strep can lead to an autoimmune reaction in which the body's immune response goes overboard and begins to attack other parts of the body. When this causes inflammation of the heart and joints, it's called rheumatic fever. When it attacks the basal ganglia, the part of the brain that controls voluntary movements, learning, and emotions, it's called Sydenham chorea. But sometimes the autoimmune inflammation in the basal ganglia appears in children who don't show signs of rheumatic fever or chorea, though they may show many symptoms attributable to inflammation of the basal ganglia, including obsessions, compulsions, tics, and ruminative thoughts. In those kids, it's called PANDAS.

In 1998, Dr. Swedo and her team described five diagnostic criteria of PANDAS, simplified as follows:

1. Presence of OCD or a tic disorder

2. Onset before puberty (this was eventually updated to include all children under the age of eighteen)

3. Symptoms begin abruptly and subsequently wax and wane

4. Patient tests positive for strep antibodies, indicating a current or previous Group A streptococcal infection, which seems to coincide with the onset of the OCD or tic disorder symptoms

5. Association with neurological abnormalities (in addition to tics, which can be overt and "violent," patients may repeat smaller involuntary movements, like outstretched "piano-playing" fingers)

I once watched strep set a patient back by two years. I had diagnosed Lief with a nutritional deficiency when he'd been brought in around age eight with stunted growth and anxiety. On a robust regimen of nutritional supplements and dietary modifications, he grew over a foot and gained enough confidence and self-esteem that when he was ten years old his parents and I agreed that he no longer needed to see me.

His mother, Nadine, wasn't alarmed when Lief came down with his first bout of strep about a year later, when he was eleven. She gave him the antibiotics prescribed by his pediatrician, waited for the fever to dissipate, and sent him back to school. Then Lief's teachers started calling. Lief was cursing in class and flying off the handle at the slightest provocation. In addition, he'd developed a stutter so severe he could hardly communicate in class, as well as incessant blinking that made it hard for him to read or focus on anything. As he realized he couldn't control the stuttering or blinking tic that accompanied it, Lief's anxiety came roaring back.

The tics and stutter were so pronounced I noticed them immediately as the boy walked through my office door. The follow-up throat culture was negative, suggesting the strep was gone, but I had little doubt the strep infection had provoked the brain inflammation causing his symptoms. It took almost a full year, but with an additional course of antibiotics (frequently these PANDAS kids need more than just a ten-day course of antibiotics), dietary changes (including removing foods containing gluten and dairy), and some immune system modulators (natural or medicinal products that help increase or decrease immune function as necessary to restore balance), the stuttering, blinking, and mood swings resolved, and the anxiety lessened.

Ultimately, to account for the multitude of patients exhibiting acute-onset PANDAS symptoms who tested negative for the S (streptococcal infections), in 2012, it was proposed that PANDAS should now be considered a subcategory of what would be called PANS—pediatric acute-onset neuropsychiatric syndrome. The new syndrome would encompass all types of triggering factors besides strep, and not just infectious diseases. The criteria are as follows, with some simplification:

1. Abrupt, dramatic onset of OCD or severely restricted food intake

2. Simultaneous presence of additional neuropsychiatric symptoms, with similarly severe and acute onset, from at least two of the following seven categories:

   i. Anxiety

   ii. Extreme mood swings and/or depression

   iii. Irritability, aggression, and/or severely oppositional behaviors

   iv. Behavioral (developmental) regression (start acting younger than their age)

   v. Sensory or motor abnormalities

   vi. Deterioration in school performance

   vii. Physical signs and symptoms, including sleep disturbances, bed-wetting, or urinary frequency

3. Symptoms are not better explained by a known neurologic or medical disorder, such as Sydenham chorea, systemic lupus erythematosus, Tourette's disorder, or others

Unfortunately, despite multiple studies supporting the diagnosis and increasing numbers of physicians around the country identifying it and treating it successfully, PANDAS still isn't universally accepted as an official condition. No one argues that previously healthy children can't fall prey to a sudden onset of intense OCD symptoms. But skeptics have questioned the data pointing to strep as their cause. Given the prevalence of strep in childhood, isn't it more likely, they wonder, that most of the acute-onset OCD kids had strep simply because most kids get strep?

It's true the supporting studies have been small, and not all have met the gold standard of medical studies—random, double blind, placebo controlled—but they are numerous and convincing enough to take seriously. Coupling the research with my own clinical experience and that of many other pediatricians and family practitioners, it's hard for me to understand how this disorder could still be dismissed as purely psychological. As more physicians become aware of the mounting scientific evidence connecting severe OCD symptoms, infection, inflammation, and autoimmunity, I'm hopeful that families will be able to find help for their children quickly and locally. It's also encouraging that several prestigious institutions, including Stanford University School of Medicine, the University of Arizona, the University of South Florida, and Massachusetts General Hospital, have established PANS research and clinical divisions to further study and treat patients with the syndrome.

Introducing ITABI

As all-encompassing as this relatively new diagnostic nomenclature is, to me it still overlooks one of the main factors of the disorder. As we've discussed, autoimmunity occurs when the immune system can no longer differentiate between self and nonself and produces antibodies that attack the body, which causes inflammation. You can have inflammation without autoimmunity, like when your body is healing from a cut or even larger trauma, but if you have autoimmunity, inflammation is always present. Given the enormous role inflammation plays in this condition, the name PANDAS, which doesn't mention inflammation and refers only to strep, seems too restrictive. And though PANS is a better name, it too, unfortunately, evokes controversy. I coined a new name that I feel better reflects this brain disorder: infection-triggered autoimmune brain inflammation, or ITABI.

It's never just one factor that contributes to inflammation and a misdirected autoimmune response, but rather an "environmental mosaic" of genetic vulnerability, immune imbalance, hormonal imbalance, environmental factors, including toxicants, and nutritional deficiencies. The Western diet, which tends to be laden with fat, sugar, salt, and processed food, can cause inflammation that weakens the immune system as well as the tight junctions in the gut. We already know that the health and balance of the microbiome and the immune system are connected, which supports research that has found an association between an imbalance in the gut, autoimmune disorders, and psychiatric disorders. And we have also learned that in individuals who find themselves living with the right combination of genetic predisposition and environmental triggers, infections—including infections as normally innocuous as strep throat, sinus infection, mycoplasma pneumonia, flu, chicken pox, herpes, and even the common cold—can have a substantial impact when they lead to inflammation, increased permeability of the blood-brain barrier, and ultimately inflammation of the central nervous system.

There are two ways in which infections can promote inflammation and autoimmunity in the brain. The first is a process called molecular mimicry, in which tiny pieces (peptides, or small proteins) of bacteria and viruses, known as epitopes, stimulate a misdirected immune response to a piece of an organ that looks just like the epitope. In the case of brain inflammation, these epitopes are identical to tiny regions of the cells of the basal ganglia. Let's say a strep pathogen enters the body. The immune system responds to the pathogen by making antibodies. An epitope of strep bacteria, however, looks just like a small piece of the basal ganglia. When those strep antibodies go hunting, they'll see small parts of basal ganglia cells that look exactly like the pathogen they're designed to eliminate. Mistaking them for the pathogen, they'll go on the attack, causing inflammation of the brain.

The other way the body triggers brain inflammation is not through antibodies but through a cell-mediated mechanism that results in a dysregulated proliferation of the "heavy hitter" inflammatory immune cells, Th17. When activated by strep and other infections, especially in the presence of a leaky blood-brain barrier and a Treg imbalance, these specific T-helper cells can breach that barrier and attack specific brain regions, including the basal ganglia, leading to autoimmune inflammation and subsequent neuropsychiatric symptoms.

My patients with ITABI are often some of the most heartbreaking cases I treat. Many of them are at the top of their game—intelligent, athletic, social, happy kids. They come from loving homes and they're close to their families. And then they wake up one day completely different. Though sudden onset is part of the classic definition of ITABI, occasionally it happens over a few days or weeks. Still, it's dramatic and fast, the behaviors and symptoms depending entirely on how or whether other parts of the brain adjoining the basal ganglia are affected by the immune system's misdirected attack. Suddenly, kids are lining things up, insisting objects need to be placed in a certain order, walking back and forth through doors five times in a row, and washing their hands so often their skin cracks and bleeds. They cut themselves, hit the walls, and scream with rage. They can't concentrate. They forget everything they've learned, exhibiting dyscalculia (the inability to calculate numbers), dysgraphia (the inability to write legibly), and dysphonia (the inability to control one's voice). They wet the bed and feel the need to urinate frequently, though there's no sign of a urinary tract infection. Their contamination phobias are so severe they can't touch anyone else or be touched, putting a terrible distance between themselves and loved ones. They can't eat for fear of what's in their food, causing severe weight loss. They have nightmares. They hallucinate and hear voices. They can be suicidal. They can be homicidal. The behavior is so violent and frightening it can tear families apart. And yet, with proper treatment, most of these children do get better, especially if their illness is caught and treated quickly. (This is critical—the longer the brain is inflamed, the more risk the neuroinflammation can become chronic, with lasting and long-term effects. However, I have helped many kids whose symptoms were present for years, so there is hope.)

This is where Georgia got lucky. Her mother's determination, and her professional connection to functional medicine practitioners, put her on the path to unlocking Georgia's medical mystery. Within a few days, a colleague who'd heard about the reason for Natalie's abrupt departure from the Florida conference put her in touch with my office. Luckily, an opening unexpectedly appeared in my schedule, and sooner than she had thought possible, Natalie had Georgia in the car, driving eight hours through a blizzard to see me in upstate New York. From the backseat, Georgia asked, "Am I in Heaven or in a dream?" so many times in a row Natalie almost managed to tune her out, until Georgia changed tack.

"Mommy, I have to tell you something."

"What, honey?"

"I would never do this. I need you to know I would never do this," said Georgia. And then, almost in slow motion, she continued, "The voices in my head are telling me to kill Danny while he's sleeping."

Natalie's mind reeled. *Oh my God oh my God oh my God!*

Mystery Solved

"I want to feel normal again," said the tiny girl with the big hazel eyes. Her mother turned to look at me, depleted from having just spent an hour and a half answering my questions and listing in excruciating detail the painful events of the last five weeks, as well as her daughter's entire medical history. I looked over my extensive notes, taken during this initial consultation. A full-term, C-section birth, formula fed, some cradle cap. Recurrent ear infections that eventually necessitated tubes. Three episodes of pneumonia. A case of hives following her chicken pox vaccination. Allergies to penicillin, amoxicillin, shellfish, peanuts, and soy. As a small child she showed mild symptoms of OCD, lining up her princess shoes in a perfect line, but was otherwise described as "the happiest kid on earth." She'd been active in sports and enjoyed playing in the woods behind her school. More recently, she would wake up in the morning not knowing where or who she was. She also suffered from air hunger, during which she would struggle to get enough oxygen. She had a fear of contamination, brushing her teeth to the point of gagging, yet resisted taking showers. She experienced frequent episodes of significant anxiety, rage, and frustration due to a feeling of helplessness and extreme separation anxiety, compounded by having a mother who had to travel a lot for work. She'd been to the pediatrician multiple times and had no pain or fever, but had tested positive on an anti-streptolysin O titer test.

That test result was consistent with my impression that Georgia's sudden, dramatic symptoms were likely caused in part by the strep infection for which she'd tested positive a few weeks earlier (confirmed by another throat culture in my office), which made her a classic case of ITABI. Unfortunately, Dr. O'Connor's assurance that Natalie had figured out the cause of her daughter's suffering was premature. No one becomes this ill from a bout of strep. They can, however, suffer neuropsychiatric disorders when genetics and other environmental factors compound or add to infections and inflammation, causing their immune kettle to overflow. ITABI is a prime example of how the interaction of environment (i.e., infections), increased gut permeability, and blood-brain barrier permeability can create a perfect storm of autoimmune response, which results in widespread inflammation. In fact, it's the disruption to the blood-brain barrier that allows the inflammatory immune molecules and cells to get into the brain and wreak their havoc.

When treating a child with an immune dysregulation, it is essential to take a comprehensive, individualized approach, evaluating and treating not just the neuroinflammation but all other factors of the immune kettle. The best way to protect these kids from future illness is to restore their health systemically. If we don't lower the levels of the immune kettle, the next pathogen, toxicant, or stressful event has the potential to make them very sick all over again.

Georgia's history of allergies and recurrent illnesses suggested that she had a weakened yet still hyperreactive immune system. One big red flag was the pneumonia—three episodes is a lot for a ten-year-old kid. Like most children, she had received a pneumococcal vaccine, specifically the 23-valent pneumococcal polysaccharide vaccine, PPSV23, designed to protect against twenty-three subtypes of strep pneumonia, which is different from the Group A beta-hemolytic streptococcus that causes strep throat. Georgia's labs showed that she was only immune to three of those twenty-three. The fact that her immune system hadn't developed an immunity to most of the subtypes was an indication of its impaired function.

I think of every office visit as a game of chess between myself and my patient's illness. Assessing the information given to me and weighing the benefit-to-risk ratio of all possible outcomes, I make a move. Then I wait to see how the patient responds. Their response tells me what my next possible move might be, and once again I evaluate the risk benefit of one course of action over another. It's slow and time consuming, but this is what individualized medicine looks like. You're processing a patient's information on top of your experience and your clinical knowledge base, allowing you to consider and evaluate all possible treatments the way a master chess player assesses a board and sees all possible permutations of the game. I was certain Georgia was suffering from ITABI, and started a plan that I hoped would deliver a checkmate.

Treatment

Lower the Inflammation

To begin, we needed to get Georgia's inflammation under control. I put her on a course of ibuprofen every eight hours for a week, as well as some natural nonpharmacologic anti-inflammatory treatments, including omega-3 essential fatty acids, probiotics, vitamin D, and curcumin (one of the most potent natural antioxidants and anti-inflammatories). Something was activating the microglia in her brain, causing them to pour out large numbers of inflammatory mediators. Until we figured out exactly what it was, hitting it with anti-inflammatories couldn't hurt. We also discussed a possible short course of the steroid prednisone.

Treat the Infection

The next step in my strategy was to attack the infection. I prescribed azithromycin, an antibiotic, to kill the underlying strep infection that was driving the misdirected immune response. Unfortunately, if an infection had burrowed deep in the tissue, which I believed is what had happened to Georgia, a seven-day course like the one she had taken back when she'd first started exhibiting symptoms of OCD wasn't sufficient. She was going to have to take them for a while, not just until the titer numbers started going down (because they can decrease very slowly), but until her behavior started to improve. There was no way to know exactly how long that might take. Kids with rheumatic fever can be on antibiotics for years, so often children with ITABI, a cousin to rheumatic fever, need longer courses, too. This makes many doctors uncomfortable due to the possible adverse effects, and a prolonged course of antibiotics is not something to be done lightly. But sometimes it's the necessary course. I suspected this was going to be one of those times.

I got a call from Natalie about two hours after she'd administered Georgia's first dose of antibiotic. Georgia was worse, much, much worse—kicking, punching, and screaming in a massive fit of rage. She kept asking for the code word, lollipop, but it wasn't working. Natalie was frantic. Was it possible she was having an allergic reaction to the antibiotic?

She wasn't. The best way to describe what Georgia was going

through was die-off. To Natalie, it probably looked terrifying. This dramatic, painful process is frequently referred to as the Jarisch-Herxheimer reaction, in the vernacular known as "herxing," a die-off reaction first reported by dermatologists who noticed that the lesions suffered by their syphilis patients got worse upon treatment before they got better. Dying bacteria release a lot of toxic by-products. Our immune systems are not just built for defense, but for repair. If you've been spending too much of your resources on defense (as Georgia was), you have less to help you with repair, which includes waste management. When you kill a massive amount of bacteria, you have to clean up the debris. If you don't, you can develop more inflammation. I prescribed a combination of activated charcoal and other nonabsorbable binders to latch on to the toxins in the gut, and help get Georgia through this miserable period.

Be Alert for Strep

As discussed previously, symptoms are not just a sore throat, but usually a combination of sore, red throats, fever, and swollen lymph glands under the jaw or on the upper neck (and sometimes even atypical symptoms such as headache, vomiting, or abdominal pain). If your child has ever had an episode of ITABI, you'll want to be extra vigilant in pursuing throat cultures and antibiotics to prevent a flare.

At the very first onset of illness, you can offer a child natural antimicrobials such as olive leaf extract and elderberry, as well as supplements that help support the immune system like transfer factors, a group of tiny peptides, usually drawn from bovine colostrum or egg yolk, that can convey their immunity to their host (we use a spray form to make it easier for kids to take). You can try IAG larch arabinogalactans, astragalus, vitamins A, C, and D, and minerals such as zinc and selenium, as well. It's actually not a bad idea for tweens and teens to take some of these supplement even before they get sick, especially in the fall and winter, to help support their immune system.

Check for Other Infections

About two weeks after Georgia's initial visit, the Lyme and co-infection test results came in. She'd tested positive for the IgM titer to a relatively uncommon tick-borne disease caused by the bacteria *Francisella tularensis*, possibly picked up while playing in the woods behind her school. IgM refers to a class of immunoglobulin, the first antibodies the body makes when it encounters a pathogen, as opposed to IgG antibodies, which are produced weeks after exposure and "remember" the illness so they can be ready to fight it the next time they see it. The *Francisella tularensis* could help explain why the strep infection affected her so severely. Occupied with the tick-borne disease, her immune system hadn't been able to effectively deal with the leftover strep. I called in a second antibiotic prescription for doxycycline, which treats tularensis more effectively than azithromycin. Natalie administered the first dose immediately. The next day was the best both mother and daughter had experienced in months. While Georgia was still asking whether she was in Heaven or in a dream, still performing rituals, still refusing to eat, and experiencing anxiety about food in general, she managed to go to school. After what Natalie and Georgia had been through, that small step forward felt like nothing short of a miracle.

Protect the Gut

Given the large doses of antibiotics Georgia was taking, we needed to be proactive about protecting her gut from the assault on its healthy bacteria. Fortunately, Natalie had been a medical consultant for over a decade and was well versed in supplements and nutritional products. Before Georgia even took her first dose of antibiotic, Natalie had her on a strict regimen of prebiotics and probiotics. Even though Georgia was on two antibiotics for several months, she never suffered any adverse GI side effects.

Introduce an Anti-Inflammatory Diet

As an infant and toddler Georgia was diagnosed with shellfish, peanut, and soy allergies. The presence of so many significant food allergies suggested that her gut health wasn't optimal and that, in fact, she may have a leaky gut. We created an anti-inflammatory nutritional protocol for Georgia that would minimize her vulnerability to disruptions in the gut-brain-immune axis. Even if she didn't have any allergies, I recommended removing gluten and casein (a protein found in most dairy products) from Georgia's diet.

You'll see that we talk a lot about gluten and casein in this book, and that in almost every case I recommend eliminating one, the other, or both from a child's diet when I suspect inflammation. That's because gluten and casein are multisystemic inflammatory agents that can affect not just the gut but organs throughout the body, including the blood-brain barrier. I learned this when working with autistic children and saw significant improvements in many children when we removed gluten and dairy from the diet. I subsequently applied this knowledge to neurotypical kids with neuroinflammation and neuropsychiatric symptoms, and saw similar types of positive results in many of them.

Sometimes simply removing casein or gluten, or even refined sugars, isn't enough, at which point I may recommend a more anti-inflammatory type diet. Essentially a modified Paleo diet, an anti-inflammatory diet avoids most grains and is high in protein, good sources of fat like avocados, olive oil, and coconut oil, and packed with vegetables and fruits. Researchers have also found that a strict ketogenic diet (one that avoids carbohydrates, including those found in starchy vegetables and fruits) can block the formation of inflammasomes, or innate immune sensors/receptors that trigger inflammation in the presence of infectious microbes or even pieces of self-proteins (which we see in autoimmunity). This may help explain at least some of the diet's anti-inflammatory effects. I generally don't prescribe ketogenic diets because they are extremely strict and hard to implement. More often, I'll recommend a modified ketogenic diet, which is similar to the Paleo diet but lower in carbohydrates, avoiding even fruits. Originally, ketogenic diets were pioneered to help control seizures that weren't responding to typical antiepileptic drugs (AED), and also showed promise in stabilizing mood disorders. However, modified ketogenic diets are not easy to follow either, and therefore I generally recommend them only to families with significantly affected children, advising they work closely with a nutritionist to successfully implement them and ensure adequate nutrient intake.

Sometimes people will think there's no harm in adopting a strict diet, reasoning that if it's good for some, it's probably good for all. These diets do have risks, including symptoms such as fatigue, diarrhea, and nutrient deficiencies. They also require more caution for a person with diabetes, kidney disease, heart disease, or gallbladder dysfunction. These stricter types of diets need to be undertaken with the help of a nutritionist to ensure a child gets a healthful diet replete with necessary nutrients. A child with mild symptoms most likely doesn't need to be on a modified ketogenic diet; simply cutting gluten or dairy, or just decreasing sugar and refined carbs, may produce improvements.

I was looking for every possible way to heal Georgia's gut wall and extinguish the inflammation making its way to her brain. But it wouldn't be enough to address the permeability of the gut; I also needed to address the permeability of the blood-brain barrier.

Shore Up the BBB

I treated Georgia with nutrients that have been shown to help decrease the abnormal BBB permeability. These included vitamin D, omega-3 essential fatty acids, B vitamins (including methyl B-12, methylfolate, and B-6), magnesium, melatonin, and a combination of antioxidant/anti-inflammatory phytonutrients consisting of resveratrol, pterostilbene, curcumin, and sulforaphane. (See appendix for more details.)

Promote Calm

One treatment that I'm using more frequently these days because it has such excellent antianxiety and anti-inflammatory properties is cannabidiol (CBD), delivered in oral drops under the tongue. Though this derivative from the hemp plant is found in recreational marijuana, it has only minimal amounts—.03 percent or less—of THC, the psychoactive component of pot, far too little to get anyone high (I use broad-spectrum CBD that contains 0 percent THC). The World Health Organization has stated that CBD is not addictive and cannot be abused, and that "there is no evidence of . . . any public health related problems associated with the use of pure CBD."

Our bodies are actually designed to bond with and break down cannabinoids. In fact, through something called the endocannabinoid system, we naturally produce our own cannabinoids, which help regulate cells in the brain, endocrine system, immune systems, and gastrointestinal tract. They also affect our stress response.

The endocannabinoid system has two main receptors, CB1 and CB2. Though these receptors are found throughout the body, CB1 is present primarily in the central nervous system's spinal cord and brain, where it affects the neurons, but also in other parts of the body, including the enteric nervous system in the GI tract. CB2 receptors are found mainly in the peripheral immune system, such as the liver, spleen, tonsils, and bone marrow, where it targets the immune cells, such as macrophages, B cells, and T cells, but also mast cells, and microglia in the brain, which may help explain its far-reaching effects on inflammation. They are also found in the peripheral nervous system.

These receptors bond with our naturally produced cannabinoids, one of which, anandamide, has been dubbed the "bliss molecule." Anandamide is responsible for the "runner's high" you can get from exercising as well as the mood boost some people experience when they eat chocolate. Anandamide levels have also been shown to rise during meditation and acupuncture. But CB1 and CB2 can also receive phytocannabinoids, the ones not produced in the body but that we ingest, such as through CBD.

CBD has been shown to have neuroprotective effects and to decrease autoimmune symptoms in the following ways:

- Reducing microglia activation
- Downregulating pro-inflammatory mediators, such as TNF-α and IL-6
- Downregulating inflammatory Th17 cells and inducing T-regs
- Stimulating synaptic plasticity, meaning it increases the brain's resilience and flexibility
- Facilitating neurogenesis, which means it increases its ability to regenerate and heal

All of these improvements—and the fact that CBD may contribute toward decreasing psychosis, anxiety, and depressive behaviors—mean it's an especially beneficial treatment to consider for my ITABI patients, and for many of my other patients as well.

Because CB1 receptors are present in so many cells throughout the body, we can see a cascade of positive effects as it influences other channels, intracellular messengers, and pathways, a synergism referred to as the "entourage effect." When used in conjunction with terpenes and flavonoids, other nonpsychoactive extracts from the cannabis and hemp plants, I've seen CBD promote better sleep, decrease headache and migraine, and reduce anxiety in many patients.

Some kids who've previously tried CBD report that they haven't fully experienced its effects until they start the regimen I prescribe. That's because dosage and quality are extremely important. The CBD industry is unregulated, which means there are a lot of products out there of questionable quality. Some of these may contain pesticides and herbicides and might only have trace amounts of CBD in them, not enough to have any significant clinical effect. I recommend a high-quality, pharmaceutical grade CBD that's been third-party tested and received a certificate of analysis confirming it's free of toxins, herbicides, and pesticides. I start my patients on a low dose and, as we see how they tolerate it and respond, work our way slowly to a higher dose until, hopefully, we see the improvements we're looking for.

Quell Autoimmune Inflammation

A month after beginning treatment, Georgia was experiencing some mild progress, but she was still, in the words of her mother, a shell of herself. In our early meetings, when I had outlined possible treatment options, I mentioned that some kids saw remarkable results with something called intravenous immunoglobulin (IVIG), a type of IV blood therapy that can help quell the autoimmune inflammation in the brain. It seemed like it might be time to give it a shot.

IVIG would generally not be my first choice for a treatment plan, and I usually reserve it for children whose symptoms are interfering with their ability to function. It's an uncomfortable procedure for kids, requiring that they remain hooked up to an IV for two full days while the antibodies derived from one thousand to fifteen thousand blood donors are pumped into their bodies to decrease the inflammation and autoimmunity in their brains. It's generally considered safe, but it does have some rare and potentially significant adverse effects, and can have a few somewhat more common unpleasant side effects such as headache, nausea, and vomiting. I explained these risks in great detail to Georgia and her mother, but as time went on and her improvements inched forward at a crawl, Georgia began asking for the IVIG treatments. Her symptoms were still torturous and she was desperate to feel like herself again. Natalie had found a number of PANDAS groups online and asked parents there about their children's experiences with IVIG. After hearing several promising reports, Natalie decided to pursue the treatment—which is not always covered by insurance and can be quite costly—in hopes that it might be just the game changer Georgia needed.

We did all the appropriate screenings before setting Georgia up for the treatment. If she hadn't had ITABI—had she been a patient with, say, a primary immune deficiency—I would have suggested a low dose of IVIG to boost her immune system. In Georgia's case, the goal was to quiet her overactive immune system and quell the inflammation that raged in her brain, which required a larger dose. Often it takes at least two to three treatments of IVIG before a patient experiences noticeable results. Fortunately for Georgia, one round was enough.

Three weeks after her treatment, Georgia was 70 to 80 percent back to herself. Whereas before I could only qualify her progress as mild, now it was definite. She was still asking her Heaven-or-in-a-dream question, still having negative and intrusive thoughts, but the fits of rage stopped. In time she was able to make it through a playdate at a friend's house. She was able to brush her teeth again. She enjoyed a session of summer camp. And with each follow-up appointment, she continued to improve. Georgia's nightmare began on March 19. By September of that same year, she was healed.

Happy Ending

If you were to meet Georgia today you'd never guess that she had once suffered such a debilitating illness. She's in therapy to help manage the remaining traces of OCD, learning to break her obsessive and compulsive habits by replacing them with positive ones. But in all other ways, she's a regular fourteen-year-old who generally sleeps through the night, goes to school, plays sports and works out, enjoys spending time with her mother, father, and brother, and craves the company of her friends. At this point, it's likely Natalie has a longer road to full recovery than Georgia. Seeing your child critically ill, and caring for them throughout that illness, is not only stressful, exhausting, and anxiety producing, but often traumatizing. Post-traumatic stress disorder is common among parents who have had to navigate such episodes, and it's important for them to get the help and support they need, too.

The last time I saw Georgia, there was no trace left of the anguished, terrified, haunted girl. Instead, she seemed confident, her smile stretched wide across her face, as she sweetly thanked me for helping her feel better.

Not every case of ITABI is as severe as Georgia's. I didn't choose to share this complex case to frighten you, but rather to show how even a serious, multifactorial case like this one can get better. While every step in the treatment protocol I've shared here was specific to this particular patient, the process I used to assess the symptoms and expose their root cause is the same one I use to heal any patient with a complicated illness. The questions I asked along the way are the same you might ask yourself. You don't have to see suicidal or homicidal ideation to consider whether a child's writing deteriorated almost overnight, or to question whether they showed any cognitive dysfunction, or to think back to when they started wetting their bed. By uncovering the parts and the patterns, we can reveal the whole. The combination of clinical presentations you can identify will dictate the profile of your child's Mood Dysregulation Spectrum and offer clues to whoever you find to help you deconstruct all the different elements affecting your child's health, so they can develop fresh options and an individualized treatment program.

COULD YOUR CHILD HAVE ITABI?

Did your child display symptoms after suffering from an infection?

Has your child tested positive for strep or another infection, including Lyme and other tick-borne coinfections?

Did symptoms appear abruptly, almost overnight? Do they include tics, anxiety, OCD with intrusive thoughts or repetitive actions or rituals, panic attacks, sadness or depression, defiance, mood swings (at times rapid cycling with aggression or violent acts), suicidal or homicidal thoughts or behaviors, and/or psychotic thoughts including auditory or visual hallucinations? While it would be highly unlikely a child would experience all of these, the sudden onset of more than one of these symptoms, especially tics and OCD, would be a strong indication of ITABI, as depicted in the MDS profiles that follow.

Does your child have a history of tics or OCD, which appeared and then seemed to wax or wane?

Has your child's behavior regressed so they act much younger than their age?

Has their handwriting suddenly become difficult to read, or have their drawing skills deteriorated?

Has your child lost their ability to do math?

Did your child start wetting the bed at night or urinating very frequently, without evidence of a urinary tract infection?

Has your child become suddenly hyperactive or lost his ability to concentrate and pay attention, or have her ADHD symptoms suddenly worsened?

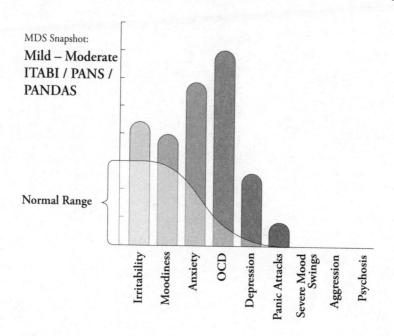

MDS Snapshot:
**Mild – Moderate
ITABI / PANS /
PANDAS**

Normal Range

Irritability
Moodiness
Anxiety
OCD
Depression
Panic Attacks
Severe Mood Swings
Aggression
Psychosis

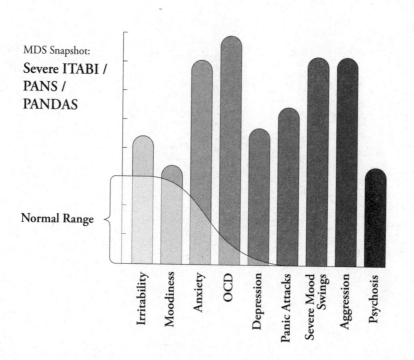

MDS Snapshot:
**Severe ITABI /
PANS /
PANDAS**

Normal Range

Irritability
Moodiness
Anxiety
OCD
Depression
Panic Attacks
Severe Mood Swings
Aggression
Psychosis

CHAPTER 5

Bitten by the Bug

Lyme, Bartonella, and
Other Tick-Borne Diseases

Nothing—not a grizzly, not a hurricane, not a fire—compares to the ferocious intensity of parents determined to help their child who is suffering. Had I not been able to relieve Georgia's symptoms, I know without a doubt Natalie would have continued to hunt for answers, no matter how long it took to get them.

Forty-five years ago, two other mothers refused to accept inconclusive diagnoses or resign their children to a lifetime of illness. It's thanks to them that we know as much as we do about a disease that remains roiled in controversy, a disease whose symptoms are so varied and mimic so many other well-known illnesses, including psychiatric disorders, that physicians may not even think to check for it.

The story began in the mid-1970s, when Judith Mensch noted that many children in Lyme, her bucolic town in rural Connecticut, were getting sick. As she told a *New York Times* reporter, first it was the little girl next door. Then it was another child around the corner. Then a boy down the street. Finally, her own daughter, Anne. Within a year and a half, all were hit with symptoms of what doctors labeled juvenile rheumatoid arthritis so severe it put them in the hospital and left them temporarily unable to walk unassisted. Just a few miles away, Polly Murray watched helplessly as first her teenage son, then his twelve-year-old brother, and finally her husband were struck with headaches, rashes, and swollen joints so painful that they struggled to walk without crutches. Polly was familiar with their suffering—she'd been hospitalized with similar afflictions for most of the fifteen years the family had lived in the area.

The symptoms did look like arthritis, but that diagnosis made no sense. Arthritis wasn't an infectious disease. How could it be spreading to so many people in one community? Each woman separately started her own investigation into the possible causes of their families' ailments, and finally called the Connecticut State Department of Health to demand answers. Not long after, a Yale team led by CDC-trained scientists took the mothers' preliminary research and started investigating the cause of what they believed was a "previously unknown disease" they called "Lyme arthritis," the symptoms of which included sudden-onset swollen joints—particularly the knees—aches, pains, and in about a quarter of the cases, a large red-and-white "bull's-eye" skin rash. In 1978 the team hypothesized that ticks were responsible for Lyme arthritis. Not long after, scientists confirmed that ticks were in fact the primary "transmission vector" for the "infectious agent" that caused Lyme disease, which four years later was identified as the bacterium *Borrelia burgdorferi*. Soon after, they recognized that the "arthritis" was merely one manifestation of a full-blown disease with symptoms that can affect the central nervous system, heart, joints, and skin. For decades Lyme disease was considered a hazard of living in the northeastern part of the United States, which is endemic to ticks thanks to its numerous wooded areas housing the mammals—especially deer—that serve as good hosts.

I only learned this history in 1985, when an adult patient presented himself to me with a markedly red and swollen toe. An allergy, I thought at first, but when my patient didn't respond to Benadryl, I knew something else had to be amiss. Grilling my patient with more questions, I found out that he had recently been on vacation on Block Island, Rhode Island. I'd never been, and I asked him what it was like. "Beautiful. Calm, serene. Lots of deer." Deer. Somewhere along the way, at a conference or in some medical journal, I was pretty sure I'd read about a disease that seemed to strike primarily in wooded areas heavily populated with deer. Could this be it? A few hours of research revealed that sure enough, some cases of a tick-borne infection called Lyme disease had been reported on Block Island. After a three-week course of the antibiotic doxycycline, my patient's toe returned to ordinary size. (These days I would treat a case like this with no less than four to six weeks of antibiotics.) In time I'd learn how lucky he was. We'd caught the disease early, shortly after he was infected, so treatment was simple and straightforward. Had he waited longer to see a doctor, or had I taken longer to connect the dots or dismissed the possibility of Lyme disease altogether—a common occurrence, even today—his swollen toe would have likely been the least of his concerns.

Years later, around 2008, I would treat a man who was suffering so badly from debilitating fatigue, brain fog, and dizziness he had to stop working. He had been tested for Lyme, but the results from his local lab didn't meet the test criteria for a positive diagnosis. That was enough for multiple doctors to dismiss the possibility of Lyme. On top of that, the man lived in Vermont, and at the time it was believed that Lyme didn't exist there. No matter that the man was a taxidermist by trade. After hearing about the hundreds of ticks he cleared off each animal skin he worked with, I ordered a new round of tests from labs with expertise in tick-borne disease, which came back positive for Lyme. It took awhile as he had been ill for quite some time, but after a course of combination antibiotics, eventually he was on his feet and back to work.

Today few would assert that Lyme can't exist in Vermont, or anywhere else for that matter. Whether it's because people are traveling more, or that climate change is making other parts of the country more hospitable to the disease-carrying parasites, or urbanization is bringing hosts like deer into closer contact with humans, the number of tick-borne diseases reported around the country was more than double in 2016 than in 2004, with Lyme representing more than 80 percent of those cases. Around 300,000 cases are diagnosed in the U.S. per year, more than 10 times the number of cases that are actually reported to the CDC. That's over 1.5 times the number of women's breast cancer diagnoses and six times higher than cases of HIV. By 2018, every state had reported incidences of Lyme. This means that even if you live in an area not commonly known for tick-borne diseases, when presented with an illness without a cause, it's wise to consider Lyme as a possibility.

Unfortunately, even living in a Lyme hotbed doesn't guarantee that doctors will seriously consider it when they can find no other explanation for an illness. Incidence of the disease had been steadily rising in Massachusetts when one of my patients, Stacey, was born there in 2000. And yet when Stacey's parents carried her across the threshold of my clinic in 2013—the girl too tired and depressed to walk after nine years of chronic insomnia, paranoia, anxiety, vicious meltdowns, and OCD—I was the first doctor who would even entertain the possibility that Lyme could be at the root of the illness that had turned her from a socially well-adjusted athlete into a debilitated recluse.

A Complex Case

In the beginning, Stacey's mother, Rita, could find an explanation for everything.

At age four, Stacey, known as an affable, curious, talkative little girl who loved school, gymnastics, soccer, and swimming, developed an intense fear of thunder and lightning that translated into panicked refusals to leave the house if there was a single cloud in the sky. Rita chalked it up to the time Stacey's grandmother hurried her from the beach as a thunderstorm approached.

Stacey's tantrums started around the time she turned six, unpredictable rages so violent Rita bolted Stacey's furniture to the wall to keep her daughter from pulling everything down on herself as she ripped through her room like a tiny Tasmanian devil. One day as she found herself once again crouching and crying on the other side of the locked door, waiting for Stacey's fury to subside, Rita remembered that kids with poor sleeping habits often develop behavioral problems. Sure enough, a visit to the doctor revealed "kissing" tonsils, so enlarged they were touching. They were surgically removed within the week, and the rages abruptly stopped.

At age seven, Stacey started showing signs of anticipatory worry before school and hypervigilance once there, nervously asking questions about what was going to happen throughout the day, as if to prepare herself for some unanticipated disaster. Rita believed her child's insecurities were being stirred up by some bullies in her second-grade class. Or maybe Stacey was just an anxious kid. But eventually this behavior, like many of the others, faded away to the point that Rita stopped worrying about it.

When Stacey was eight, a new concern emerged. Whenever she was in the company of her best friend, Wendy, or her little brother, Derek, Stacey became obsessively concerned they were going to vomit. Her questions were relentless. "Do you feel okay? Are you going to throw up?" Rita thought it was likely due to an unpleasant memory of a long drive through Colorado that ended with her brother getting sick in the backseat of the car. Rita started taking Stacey to a therapist, who believed they were dealing with run-of-the-mill OCD. Rita liked this doctor very much, but after six months this professional with twenty-five years of experience advised her to pursue other therapeutic approaches, as none of her usually effective treatments were working.

When Stacey turned nine, Rita ran out of explanations. Aside from multiple sinus infections, Stacey was having a stellar year. Rita could come up with no reason why the girl, who adored her new fourth-grade teacher and was thrilled to have her best friend in her class, suddenly refused to ride the school bus and then started hiding on the floor of the car to avoid getting out to go to school. When Rita or her husband would ask their daughter to tell them what was wrong, she couldn't. "I don't know. I'm just so scared!" Rita had no answer for the guidance counselor who called because Stacey was crouching in terror in the corner of her office, refusing to go back to class. Despite the years of odd fears and behaviors, until now Stacey had functioned like any other young child who enjoyed school, sports, and the company of her friends. Those days were about to come to an end.

Before she was a mom, Rita had worked as a nurse for a neurologist at a major children's hospital. Whatever was going on in Stacey's brain, she was sure he could get to the bottom of it. A full neurological exam revealed nothing unusual, but the neurologist diagnosed Stacey with anxiety, ADD, separation anxiety, and a nasal tic that Rita hadn't even noticed yet. He prescribed a stimulant to manage the ADD and an SSRI (a class of antidepressants) to ease the anxiety, and suggested Rita keep Stacey out of school for two weeks until she was feeling calmer.

Far from calming Stacey down, the medicine was like jet fuel to her brain. The girl was in constant motion and her bouts of rage returned, along with explosive bursts of full-throated screaming. Rita took Stacey back to the neurologist multiple times. Convinced all he needed to do was find the right combination of drugs, he treated Stacey with a barrage of different meds and doses. Some turned Stacey into a zombie, others had no effect whatsoever or seemed to exacerbate her mood swings. She didn't return to school when her two-week break was up; there was no way she could function in a classroom. She stopped playing soccer and swimming in the family pool. She adopted new compulsions, too. Now she was unable to walk up or down a step without first sitting on her bottom and twirling around. She had a compulsive leg drag. At home, she was stepping over invisible walls. And without question, if there was a crack in the floor or sidewalk, she had to step over it.

One day as they were driving home from running errands, Stacey, sitting in the backseat, asked if Rita could make a detour to Petco to see the cats up for adoption. It wasn't a first; visiting the cats had become another compulsion, one that Rita indulged when she was able. There really wasn't time today, Rita explained. She still had things to do at home before getting dinner ready. The next thing Rita knew, Stacey was wildly kicking the back of her seat. It was everything Rita could do to stay calm and drive straight. When Rita pulled into the driveway and parked the car, Stacey got out and started throwing rocks at the house windows. Her screams echoed through the quiet neighborhood. "I hate you, Mommy!" Rita wondered if the neighbors would call the police.

At the recommendation of her pediatrician, Rita and her mother-in-law took Stacey to the ER that night, where they spent the next fourteen hours in a room watching Stacey cry. After hours of tests and interrogations to make sure Stacey wasn't being abused or otherwise traumatized by her parents, Rita was told by the attending physician there was nothing the hospital could do and that she needed to take her daughter home. Fearing what might happen when they left, Rita pleaded with the doctor to keep Stacey safe in the hospital for observation overnight. However, she didn't meet the hospital's criteria for psychiatric admission, and they couldn't hold her. Rita gripped the steering wheel the whole forty-five-minute ride back to the house, afraid her daughter would jump out of the car onto the highway. But Stacey was exhausted and spent most of the drive sleeping or staring vacantly out the window.

When Rita had spoken to the pediatrician at the beginning of Stacey's episode, he'd said, "You know, there's this thing called PANDAS. A lot of doctors don't believe in it. I don't know how to treat it, but I'm wondering if she has it." In the days following the hospital ordeal, Rita stayed up late at night reading about the disorder. With the exception of the lack of sudden onset, all the symptoms were a match. Eventually her research led her to a pediatric neurologist in a nearby state with a reputation for helping children with PANDAS. In conjunction with Stacey's lab work, which found that her strep titers were through the roof, and Rita's recitation of her daughter's symptoms, especially the leg drag, he pronounced Stacey "100 percent a PANDAS kid." After a three-hour exam, during which he noted that Stacey's pupils were dilated, his first course of action was to prescribe ibuprofen and the steroid prednisone to tame the inflammation that was likely the result of a misdirected immune response related to strep. He also prescribed an antibiotic Stacey had taken before, though not in combination with the two other medications.

Two days later, back home, Stacey started yelling. But this wasn't the angry scream to which Rita had become all too accustomed. She sounded . . . happy.

"Mom, Mom, look at this!"

Rita followed her daughter's voice into the living room and watched her gleefully run up and down the stairs like a mountain goat. The compulsion to sit and scoot on her rear was gone. So was the leg drag.

For six months, Stacey was on and off antibiotics and the ibuprofen. Yet while she saw vast improvements, they never seemed to stick. The OCD had kept Stacey out of school for more than a year by now; she kept up her studies with a series of at-home tutors. Rita was anxious to help her daughter get her old life back. She decided to seek out a second opinion and brought Stacey to a well-known child psychiatrist who diagnosed her with severe ADHD and gave her two new prescriptions, Klonopin, an antianxiety medication, and Adderall, a common medication for ADHD.

Three days later, Rita and Stacey were visiting the shopping mall with Stacey's best friend, Wendy. Somewhere between the Cinnabon and Build-A-Bear, Stacey scurried up to Rita and leaned in to whisper in her ear, "I have to tell you something. I feel like I want to hurt her."

Rita didn't understand. "What?"

"It's just a thought that keeps coming into my head. I feel like I wanna hurt her, but she's my best friend. Like, it's really weird. It's driving me nuts."

Unnerved, Rita decided it was time to go home.

That afternoon, Stacey and Wendy were playing in the pool until Stacey hauled herself out and came to stand next to her mother, dripping wet, her face plastered with fear.

"Mom, I feel like I want to drown her and I don't know why. I want to hold her under water."

A visit to the ER was followed by a consultation with a psychiatrist who suspected Stacey's negative, intrusive thoughts could be an adverse reaction to the Klonopin and Adderall, and advised she stop taking them immediately. Thankfully, he was correct, and those particular symptoms abated.

Finally, the pediatric neurologist suggested that maybe a round of IVIG (the intravenous immunoglobulin therapy that worked so well for Georgia) would allow Stacey to reboot her immune system so it would have the strength to once and for all get rid of whatever was causing the inflammation that was triggering her symptoms. Stacey was already doing better on the ibuprofen and antibiotics, but her response to the IVIG was miraculous. Over the course of the two-day, eight-hour treatment, the girl lay back in her chair with a juice box at her side and a plastic IV tube attached to her arm. Rita watched her daughter's face relax by the end of the first day. "I'll never forget it. This peace came over her." For about ten weeks, Stacey was able to have an almost normal life. She reconnected with friends. She smiled. Unfortunately, after about ten weeks, her OCD symptoms resurfaced. Reluctantly, the pediatric neurologist ordered another round of IVIG. Again Stacey's symptoms settled down. The improvement was so significant Rita registered Stacey for school, where she was now a seventh grader. Ultimately, Stacey did six more rounds of IVIG at the nearby medical center, which was tough on her because she would develop debilitating migraine headaches after each treatment, which kept her curled up on the couch for days. When she could, she attended a new school, an alternative educational setting for kids with issues such as ADHD, high-functioning Asperger's, and emotional issues. She'd been happy there, but as she began to feel better following the IVIG, she became more aware of the differences between her and the other kids and started feeling uncomfortable, like she didn't belong. Then a strep outbreak sent all her symptoms into overdrive again.

Rita found comfort and connection within a broader community of PANDAS families in a Facebook group where they could trade information, share stories, offer one another hope, and lean on one another for understanding and support. Rita frequently posted about her frustration that IVIG seemed to work only temporarily on Stacey, until one day a random mother made a comment: "If IVIG is not holding, your daughter has Lyme." That wasn't possible, Rita replied. Stacey had been tested for Lyme; the results were negative.

"Generic testing stinks for Lyme," the mother wrote.

Rita resumed her research on Lyme disease. She discovered there were two highly polarized schools of thought regarding Lyme disease, a controversy so contentious it has been referred to as the "Lyme wars." One camp follows guidelines established by the Infectious Diseases Society of America (IDSA), which asserts that Lyme disease is effectively treated after one short course of antibiotics usually lasting two to four weeks. These practitioners don't believe it can be a chronic, persistent condition. Whatever causes 10 to 20 percent of Lyme disease patients to experience symptoms such as fatigue, musculoskeletal pain, and cognitive difficulties for months and sometimes years after initial antibiotic treatment, it's not Lyme disease. That said, they're not sure what else could cause what they call post-treatment Lyme disease syndrome (PTLDS). Maybe autoimmunity, maybe trauma from being sick, but it couldn't be Lyme disease because they believe there is no laboratory evidence that *Borrelia burgdorferi* can survive antibiotic treatment. And, they say, many of these patients' labs never showed evidence of a Lyme disease infection in the first place.

The most common test, and the only one recommended by the CDC, is a two-step process that measures the presence of Lyme antibodies in the blood, not the presence of the bacteria itself. Two to four weeks following a bite from a *Borrelia burgdorferi*–carrying tick, the body will see an initial rise in the immunoglobulin (aka antibody) IgM. IgM can last for about two months to ten weeks. As it decreases in the bloodstream, there's a rise in IgG, a different antibody. The antibodies merely show exposure—they don't tell you if the infection is still active or not. A purist might say that the presence of IgM shows recent exposure, and an IgG antibody shows a history of exposure. Regardless, if the first step, the enzyme-linked immunosorbent assay (ELISA) test, comes back positive for either antibody, the sample is sent through a second step, the western blot, to confirm the diagnosis. If the ELISA results are negative, no further testing is ordered. ELISA tests are considered so sensitive that while false positives are possible, hence the confirmatory western blot, a negative result is indisputable. Doctors following this school of thought will generally not diagnose Lyme disease without noting the presence or history of a rash *and* two positive lab results.

In the other camp are doctors whose clinical experience leads them to believe Lyme disease can be present even without a rash, and that it can be a persistent condition when the initial round of antibiotics is not enough to eliminate the bacteria. If all evidence other than the lab work points to Lyme disease, these doctors believe it's reasonable to consider continuing oral, intramuscular, or intravenous antibiotic treatment until all symptoms disappear, so long as the patient is made aware of the possible harmful side effects of long-term antibiotics, such as yeast infections, stomach or intestinal upset, including *C. diff* (antibiotic-induced diarrhea), or adverse effects on the liver, kidneys, gallbladder, or other organs. These doctors, supported by the professional medical organization the International Lyme and Associated Diseases Society (ILADS), advocacy groups, and their patients, report seeing symptoms improve significantly after higher-dose, longer-term antibiotic treatment.

The head of Infectious Diseases at the teaching hospital where Rita's daughter was being treated—the doctor in charge of Stacey's case—was a skeptic, insisting no literature supported the existence of persistent Lyme disease. "She's been tested for Lyme," he kindly reminded Rita. "The tests go to the Mayo Clinic. They're the best. If they say there's no Lyme, there's no Lyme."

Rita was torn. This was one of the state's top pediatric infectious disease specialists telling her there was no way this path could lead to answers. But she also knew there was research coming from the U.S. and numerous other countries suggesting that *Borrelia burgdorferi* could linger in the body and cause long-term symptoms even after treatment. She went back to Facebook. "What do I do?" she asked.

"Find a Lyme doctor," the mother replied.

Following the Clues

I learned about the exceptional clinician William Osler as a medical student at the University of Rochester School of Medicine. One of the four founders of the Johns Hopkins School of Medicine, he's renowned for introducing the concept of the medical residency program, and for pulling medical students out of the lecture hall and into the patient's room for "bedside teaching," where they could learn the art of clinical observation and how to take an in-depth history. In the early twentieth century, when Osler was a practicing internist and professor, doctors were confounded by a terrible disease that seemed able to present almost every symptom of almost every illness known to humankind—syphilis. Symptoms ranged from sores, rashes, fatigue, hair loss, low-grade fever, swollen glands, headache, and muscle aches in the early to secondary stages of the illness to lesions, tumors, blindness, neurological symptoms such as weakness, numbness, and seizures, psychiatric symptoms such as irritability and depression, and even psychosis in the later ones. With symptoms waxing and waning, potentially affecting almost any organ in the body, people frequently had no idea they were carrying the sexually transmitted disease. Its ability to mimic everything from autoimmune disease to cancer to heart disease to neurological disease to almost every psychiatric disorder, especially in its advanced stages, led syphilis to be referred to as "The Great Imitator" or "The Great Masquerader." Dr. Osler famously pronounced, "Know syphilis in all its manifestations and relations, and all other things clinical will be added unto you," or more colloquially, "To know syphilis is to know medicine." From the turn of the twentieth century until the 1940s, when it was discovered that penicillin could cure the disease, absent of easy or obvious explanations for a patient's symptoms, especially neuropsychiatric ones, physicians simply had to stay open to the possibility of a syphilis diagnosis.

Lyme disease is our next Great Masquerader. Though not a sexually transmitted disease, it is possibly even easier to contract. And like syphilis it's caused by a spirochete, a long, skinny, spiral-shaped bacteria that can change its form to evade detection and destruction. Also like syphilis, symptoms of Lyme can wax, wane, and overlap. In the early stages it can present no noticeable symptoms, or it can appear purely neuropsychiatric, especially in later stages. When symptoms do reveal themselves, they're so varied that the disease is frequently mistaken for other illnesses, such as multiple sclerosis, rheumatoid arthritis, fibromyalgia, chronic fatigue syndrome, or autoimmune disorders, and can go misdiagnosed for years. And now, with rates of Lyme disease rising across the United States, absent of easy or obvious explanations for a patient's symptoms, especially neuropsychiatric, many physicians have come to recognize that they simply have to stay open to the possibility of a Lyme disease diagnosis. At least this physician does. It brings to mind another Oslerism: "There is no more difficult art to acquire than the art of observation." Many of my patients would never have gotten better under my care if I'd based treatment solely on their lab results instead of studying that data in conjunction with what my eyes and ears were telling me.

So what did I see and hear the day Stacey's mother and father practically carried her into my office for the first time?

I saw a brown-haired, brown-eyed, thirteen-year-old girl exhausted and demoralized from years of treatments for a multitude of recurring neuropsychiatric symptoms with only partial or temporary relief. She looked beaten down from the anxiety and sadness that had led her to retreat from her favorite activities and lose her friends.

At five foot two, 169 pounds, my new patient was significantly overweight. When Rita expressed concern to Stacey's previous doctors about her daughter's weight gain, they'd dismissed her. One even suggested that the girl was secretly bingeing. Rita knew this wasn't true, Stacey was rarely out of her sight. It didn't sound like a plausible explanation to me, either. This was a girl who loved sports, especially gymnastics. I'd heard of athletes bingeing and purging to keep their weight down, but it didn't make any sense that Stacey would sabotage her favorite activity, one of the few pleasures she had left when she could muster the strength to participate. However, Stacey was currently on Augmentin, an antibiotic, Intuniv for tics, the antidepressant Sertraline, and Seroquel, an antipsychotic. Since the time she was around eight or nine, she'd received prescriptions for Ritalin, Focalin, Strattera, Lexapro, Venlafaxine, Lamictal, prednisone, Zyprexa, Zoloft, and Ativan. SSRIs like sertraline and Seroquel, which was prescribed to try to control her rapid mood swings, were notorious for side effects that included weight gain. I thought this was a much more likely explanation. Aside from excess weight, her main symptoms were severe anxiety, OCD, chronic sinusitis, mood disorder, ADHD, and tics in which she would thrust her arms forward with a jerky movement, though they seemed to be well controlled with the Intuniv. She also had a coated tongue, bloating, and gas.

I saw a very sweet, intelligent girl in obvious psychiatric distress. Unlike many of my teen patients who sometimes just sit and grunt at me, Stacey was easy to engage. Though soft-spoken, she had a strong memory and could answer my questions in exacting detail. I was actually surprised by how much she was willing to share, not just about how she felt physically, but about the whole ordeal. She was acutely aware of what she'd lost, and her fear that she may not be able to get any of it back was palpable.

Her family history held a few clues. Rita suffered postpartum depression following Stacey's birth and had a history of anxiety. Also on Stacey's maternal side, there was a grandmother with Parkinson's and a grandfather with OCD. Stacy's father had been diagnosed with ADHD. On the rest of her paternal side, there was an uncle with Guillain-Barré syndrome; mild OCD, anxiety, and Hashimoto's thyroiditis (the autoimmune thyroid disease we read about earlier) in a grandmother; and very high social anxiety and possible agoraphobia in a grandfather. Several other paternal relatives also showed signs of anxiety. Stacey had a younger brother, Derek, with a mild case of ADHD.

Given this genetic background, Stacey's immune kettle was probably more than a quarter full the day she came into the world. Under normal circumstances, she'd already have a higher predisposition for neuropsychiatric disorders like OCD and anxiety than the average person; add in exposure to an illness that could potentially inflame her brain, and it would be wholly unsurprising for these heavy-duty psych symptoms to appear.

Next, I moved on to the case history, asking Rita to walk me through the time line and details of Stacey's illness. Rita began telling me her story, beginning, as she always did, by mentioning that Stacey had developed an absolute terror of thunder and lightning around the time she was four years old.

That piqued my curiosity. Why would a child develop such a strong phobia out of the blue? Rita thought it was related to an afternoon when the child's grandmother had been forced to rush her off the beach in advance of a severe storm. But I'd once had an eleven-year-old patient who'd come to me suffering from similarly debilitating anxiety, insomnia, and panic attacks, and who'd tested positive for Lyme disease despite no physical symptoms. The child's psych symptoms eventually resolved after antibiotics, including one delivered by IV.

I started connecting the dots. Stacey's family lived in Massachusetts, a state endemic for ticks, and when she was younger she had loved to play soccer. Soccer is played on a flat field, but go to any match and you'll see that sometimes when the ball goes out of bounds, a kid has to clamber through untended bushes, tall grasses, and trees to get it back. Did Rita remember Stacey ever having a tick bite? Rita nodded. Yes, she'd found ticks on Stacey's head when a toddler, not long before the phobia began. No, she'd never noticed a bull's-eye rash.

Sneaky Symptoms

The most common sign of Lyme disease infection is the classic erythema migrans, the large, expanding, red-and-white bull's-eye rash. There's no question—if the rash appears, it's Lyme. The rash is pathognomonic, meaning a particular sign or symptom goes hand in hand with a specific disease. Yet though the multitude of graphic photos of the rash on the internet make it look like it's always enormous and distinctive, the marking can be surprisingly easy to miss. It doesn't itch, isn't raised, and is frequently not perfectly round. Sometimes it lacks the white "ring." It's not at all unusual for Lyme patients to report they never saw the bull's-eye. Sources differ on the matter, with some claiming as many as 80 percent of patients recall a rash, and others saying fewer than 50 percent do. In my experience, up to half of children and teens formally diagnosed with the disease do not report a history of a tick bite or the rash.

Other common symptoms of Lyme disease exist. "Summer flus" are sometimes not flus at all—about 50 percent of Lyme patients will exhibit flu-like symptoms. The majority of Lyme disease patients suffer from extreme fatigue. In midstage Lyme, patients can experience joint pain that's frequently migratory, as well as swollen knees, stiff neck, Bell's palsy (in which nerves that control the muscles in the face become weak or paralyzed, causing the affected side of the face to droop), headache, heart block, heart palpitations, and encephalomyelitis (inflammation of the brain and spinal cord). Finally, in late-stage Lyme, which can occur anywhere from 6 months to 3.5 years from the initial infected tick bite, patients can report waxing and waning symptoms such as abnormalities in the musculoskeletal system and both the central and peripheral neurologic systems, brain fog, cognitive dysfunction, late axonal neuropathy (paralysis and a loss of reflexes), as well as a continuation of severe fatigue, frequent and intense headaches, short-term memory problems, word-finding difficulties, cardiac symptoms, palpitations, eye symptoms, joint aches, muscle aches, fibromyalgia, and arthritis.

It's believed that if you catch the disease at its early stage and treat the patient with the correct course of antibiotics, the disease can be cured and patients will report no residual symptoms. But sometimes patients and their physicians mistake tick bites for spider bites. Sometimes the rash doesn't appear within days of the bite, and can take as long as a month to appear, so people don't make the connection between the rash and the bite. Sometimes Lyme disease tests run at this early stage reflect a false negative because it can take four to six weeks for antibodies to form. For all these reasons, treatment can be delayed.

The risk of a misdiagnosis is perhaps even more likely in children, because while neuropsychiatric symptoms often accompany the physical symptoms of Lyme disease in adults, they can be the *only* sign of Lyme disease in kids: irritability, personality change, depression, brain fog, mood disorders, school phobia, insomnia, and anxiety.

The fact that Stacey had already tested negative for Lyme was no reason for me to discount the possibility that she had it, especially with a panic disorder cropping up so soon after a tick bite.

More Clues

Rita went on, describing Stacey's intense mood swings at age six, which disappeared after doctors removed the enlarged tonsils that were probably obstructing her airway when she slept. That, too, caught my attention. Children who get insufficient restful sleep often exhibit meltdowns and rages during the day, but I agreed with the pediatric neurologist who suspected Stacey had PANDAS, which you'll remember I prefer to call infection-triggered autoimmune brain inflammation (ITABI), which can cause inexplicable rages that transform a child's personality overnight. An additional infection on top of Lyme disease could cause a kid's immune system to overboil and cause brain inflammation, and before her tonsillectomy, Stacey had frequently tested positive for strep. As previously discussed, ITABI is often preceded by strep throat. She also had a history of sinus infections, sometimes so frequent that she was treated with antibiotics several times a month. Further, both the strep and the sinusitis were treated with antibiotics. If indeed Stacey had gotten Lyme disease from tick bites when she was four years old, could the regular administration of antibiotics over the years have kept the disease under control? Could they have quelled the symptoms, with the exception of the phobia, until a final strep infection overwhelmed her immune system and caused inflammation, resulting in uncontrollable rages that disappeared only when the tonsils, and the lingering strep hiding in them, were removed?

Shape-Shifting Persisters

But then, if the strep infection causing the inflammation was gone, why would the rages return a few years later? One reason could be that the bacteria that causes Lyme disease has a waxing and waning life cycle that triggers certain symptoms during an early stage that go away, then triggers different symptoms at a later stage, and then finally manifests as neurological symptoms. Another possibility was that Stacey had contracted a new infection that triggered an autoimmune response in which the immune system attacked and inflamed her brain. Finally, there was the chance that she'd been carrying the bacteria all along, kept in check by her immune system, and something—a stressor, a new infection—

had weakened Stacey's immune system so that the bacteria could reactivate.

Borrelia spirochetes have an uncanny ability to change form when they perceive their environment as hostile. For example, upon detecting antibiotics or antibodies, they can morph from long spiral shapes to become cysts or round bodies. Antibiotics like doxycycline, the antibiotic most commonly used in Lyme disease, work by inhibiting bacterial protein synthesis, thereby stopping the multiplication of the Lyme spirochete. Other common antibiotics for Lyme disease such as amoxicillin or Ceftin kill the bacteria by disrupting its cell wall. But when *Borrelia burgdorferi* morph into their round shape, they tuck their cell wall and flagella within the intact outer membranes, thus making the typical antibiotics used for Lyme disease ineffective. Thanks to their incredible survival techniques, nonspirochetal forms of the *Borrelia burgdorferi* can be categorized as "persisters." Once the environment becomes hospitable again—in this case, if antibiotics are ceased—the bacteria can revert to their spiral shapes, and symptoms that might have been quieted return. In the spirochete form, borrelia can also invade cells, thereby becoming resistant to cell wall antibiotics and only sensitive to antibiotics that can enter the cell. They can also coat themselves with the cell's actual membrane, in essence donning a cloak that allows them to camouflage themselves so the immune system can't see them to get rid of them. Additionally, the borrelia can switch from a motile phase (able to spontaneously move on its own) to a more stationary one, which allows it to produce and embed itself within a slimy biofilm that offers similar protection from the immune system and antibiotics as when it invades the cell. *Borrelia burgdorferi* is so sneaky, so resilient, you almost have to admire it for its ability to adapt and survive.

Signs of Brain Inflammation

Stacey's elementary school years were punctuated by hypervigilance, school phobia, paranoia, fear of contamination, and uncontrollable rages. Together with neurological symptoms, like the foot drag and the tics, and the OCD, all the symptoms and behaviors screamed brain inflammation to me. Here are some of the other clues I observed:

- Chronic infections: In addition to multiple strep infections, Stacey had battled chronic sinusitis since early childhood. So many infections suggested a compromised immune system.
- Food allergies: At some point, a doctor suggested that Stacey had a sensitivity to dairy and possibly egg whites, and sure enough her symptoms improved when she eliminated these foods, as well as gluten, from her diet (though she strayed from this diet once she felt better). Food sensitivities would point to the possibility of a leaky gut, which can cause widespread inflammation. If her symptoms improved once her gut health improved, that would suggest the gut could be a contributing factor in her neurological and psychological symptoms. For as we've read, a leaky gut can contribute to a leaky blood-brain barrier, which would put a patient at a higher than average risk for brain inflammation. Increased permeability of the blood-brain barrier would make it easier for inflammatory elements, such as those triggered by Lyme disease, to get into the brain.
- Stalled progress: The temporary nature of Stacey's improvement after IVIG told me that something was blocking the treatment's efficacy. Could it be that she wasn't getting a high enough dose? Typically, low doses of IVIG are used to treat immune deficiency conditions, whereas neurological auto-immunity and brain inflammation require higher doses. Stacey's first two IVIG treatments, the ones that had worked best, were administered at a high dosage, but perhaps to mitigate her severe headaches, her doctor had lowered the dose for the subsequent ones.
- Dilated eyes: Stacey's pupils were massive the first time I saw her. Living with chronic anxiety means living in a perpetual

state of fight or flight. The sympathetic nervous system is on high alert, and to me, that suggests brain inflammation, as dilated pupils are frequently seen during flares of ITABI.

- Homicidal thoughts: Though Lyme patients typically report crushing exhaustion, there have been instances of people with Lyme committing violence against others or reporting homicidal or suicidal thoughts.

Coinfections

While ticks are the primary vector for Lyme disease, they frequently carry many other diseases as well, all of which can be transmitted at the same time. In fact, slightly over 50 percent of Lyme disease patients are also diagnosed with at least one coinfection such as babesia, mycoplasma, or Bartonella. Lyme disease can be bad; Bartonella can be worse. The same bacteria that causes cat-scratch fever, in which the lymph vessels become inflamed, Bartonella is treated relatively effectively with intracellular antibiotics when caught early. Unfortunately, because it can burrow into red blood cells, it doesn't always show up in bloodwork. When patients come in with symptoms like numbness or tingling feet, a doctor might test for Lyme, but if the Lyme titer is negative, they'll assume that whatever is ailing the person can't be tick borne. However, if they haven't checked the western blot, and they haven't tested for any of the common coinfections that travel with Lyme disease (which is especially important in patients who come from endemic areas, work with animals, or spend a lot of time outside), it's premature to rule out tick-borne diseases. And early detection is important; as with Lyme, the longer Bartonella is left in the bloodstream, the harder it is to eliminate.

The most distinguishing symptoms of Bartonella are soreness on the bottom of the feet, especially the heel; ice pick–like pain in and around the eyes; headaches; and violaceous striae, reddish-purplish tracks that look like stretch marks. But it can also cause intense joint pain, GERD, difficulty swallowing, and crawling, burning sensations in the skin. Another symptom of Bartonella is intense, violent anger, sometimes accompanied by homicidal thoughts.

The other coinfections that frequently travel with Lyme have overlapping symptoms. Babesia, however, is a protozoan parasite that invades the brain and nervous system and causes fever, chills, sweats, chest pain, and most notably, air hunger. Mycoplasma, a type of bacteria, can cause severe fatigue, dry cough, and generalized pain. The more coinfections you have, the more complicated it becomes to treat the Lyme disease.

Stacey's explosive reaction to the ADHD and psych meds could have been caused by a new Bartonella infection if she'd been bitten by another tick. More likely, however, the cause was brain inflammation, which is often what I find in children who don't respond at all or have adverse reactions to psychiatric meds. ADHD medicines increase dopamine levels. For a kid with brain inflammation, that's like throwing a lighted match on a vat of gasoline. Some psychotropic meds are actually anti-inflammatory, which may be one of the mechanisms for their effectiveness in certain patients. Considered along with all of her history, Stacey's reaction to these medications seemed extremely relevant to me.

If she did have brain inflammation, I had to consider that she could also have an autoimmune problem. Some doctors will say that the symptoms of post-Lyme are residual, that like PANDAS and strep, Lyme disease can provoke a secondary autoimmune problem that persists even when the bacterial infection is gone. That's the big controversy when discussing persistent Lyme disease: is it a chronic infection, is it an infection with autoimmunity, or is it just a case of autoimmunity?

I think it's a combination. You can kill the infection and still be left with fragments of dead bacteria that can stimulate the immune system and cause an autoimmune reaction, and at the same time bacteria can instigate molecular mimicry that triggers autoimmunity in genetically susceptible individuals. In fact, about half of patients with persistent Lyme symptoms, in particular neuropsychiatric symptoms, test positive for antineuronal antibodies—antibodies produced against nerve cells, especially when they break down and release their neural antigens. The presence of these antibodies points to evidence of autoimmune inflammation in the nervous system.

Over several decades, researchers have found emerging links between autoimmune disorders and neuropsychiatric disorders, prompting more investigation into whether a subset of what we currently diagnose as primary psychiatric disorders are in fact auto-immune disorders with psychiatric symptoms. If I ran a Cunningham panel—a special test that measures the concentration of antibodies present in the blood—and the results revealed elevated antineuronal antibodies, it could indicate that Stacey's neuropsychiatric symptoms were being caused by an infection-triggered autoimmune reaction, which would also explain why IVIG had helped calm her neuropsych symptoms so dramatically.

Testing for Answers

The first order of business in Stacey's case was to order a new Lyme test. The criteria many doctors use to diagnose Lyme disease today (presence of the rash plus two positive hits for antibodies on the two-tier test) was established by the CDC for public surveillance purposes only. It was simply a method of tracking where the disease showed up, and never intended to be used for a clinical diagnosis. Unfortunately, for lack of anything else to work with, the medical es-tablishment adapted the criteria for diagnostic purposes, resulting in huge numbers of missed cases. The labs with expertise in tick-borne disease that I work with, however, have an expanded criteria—noting a positive result even if fewer than the minimum number of bands traditionally considered positive for Lyme are detected.

For Stacey, I ordered a Lyme ELISA test in conjunction with a western blot, although nowadays I frequently order a C6 peptide Lyme ELISA and an immunoblot, which are more sensitive and specific than other ELISA and western blot tests. It's imprudent to solely rely on the ELISA, as I've seen too many people come back negative on an ELISA and yet positive on a western blot or immunoblot, and then got better when I treated them for Lyme disease. If I have a patient with a history of tick bites, or who complains of fatigue, headaches, joint aches, cognitive dysfunction, or neuropathies, but their Lyme western blot only shows four positive bands, and the CDC surveillance criteria demands a positive hit on five bands, I'm likely going to treat for Lyme disease anyway, especially if one or two of those bands are very specific to Lyme. After all, lab work is important for confirmation of a diagnosis, but it's essential to consider the whole picture.

When Stacey's results came back, they were initially negative for tick-borne disease. But she did have positive strep antibodies and evidence of autoimmunity, including antibodies to cerebral folate receptors, vitamin D deficiency, zinc deficiency, and evidence of reactivated Epstein-Barr virus. A RAST test revealed an allergy to cats and pollen. She showed evidence of borderline low thyroid function with increased thyroid antibodies consistent with Hashimoto's thyroiditis and functional hypothyroidism, which means technically her thyroid levels were borderline low, but not low enough for an official diagnosis of hypothyroidism.

Treatment

Right away, we started working to lower Stacey's immune kettle levels as low as possible, tackling her allergies and diet. Due to the coated tongue and bloating and gas, I also suspected she had oral candidiasis, or thrush—a yeast infection of the mouth—as well as fungal dysbiosis, a yeast overgrowth in the intestines that's extremely common in patients who take antibiotics, and treated her for these as well. None of these treatments addressed our top priority—the OCD, tics, and mood disorder—but by getting them out of the way as much as possible, we freed Stacey's immune system to concentrate on addressing the main source of her inflammation. Think of the body as a battlefield, and the immune system as a military with limited resources. The faster we could eliminate a smaller threat on one end, like an allergy, the faster Stacey's body could redirect contingents to add firepower where she needed it most. To that end we devised the following plan of attack:

1. To decrease allergic-related inflammation, we implemented environmental controls such as an air purifier in her room and made sure she avoided the cats and kept them out of her bedroom. When pollen counts were high, Zyrtec was helpful.

2. We put her on a low-carb, yeast-free, gluten-free, and dairy-free diet, limiting the sugars that would adversely affect her ability to fight infection and autoimmunity, make her fungal dysbiosis worse, and had likely accelerated her drug-induced weight gain as well. When Stacey had committed to her diet a few years earlier she had shown great willpower, and she readjusted to her new vegetable-heavy, low-carb, high-protein diet easily.

3. I prescribed nystatin, a swish-and-swallow medicine, to combat the candidiasis in her mouth, as well as to help deal with her fungal dysbiosis. At one point while she was on antibiotics, I needed to treat her with a more systemic antifungal, fluconazole, which provided further help in addressing her candida-related issues.

4. I started her on thyroid-supportive nutrients and herbs, though later on in the course of her treatment they weren't enough, so I needed to move to prescription thyroid medication.

All of these treatments helped us make headway in restoring her immune kettle balance, but Stacey was still terribly sick with symptoms that strongly resembled tick-borne disease. The fact that she lived in an endemic area only strengthened my suspicion. And given the lack of test results, suspicion was all it was at the time, until I finally saw something that allowed me to make a clinical diagnosis. I hadn't asked Stacey to undress during her initial visit so I could give her an all-over physical exam, believing it was important to earn her trust first. After we'd started seeing some improvement and had built a good rapport, she allowed me to do a physical exam, and it was then that I spotted the classic purple striae of Bartonella on her inner thighs. This time I sent her bloodwork to a lab that specialized in Bartonella, and sure enough, the test confirmed the clinical diagnosis.

Now I had my work cut out for me. Tackling these infections, along with Stacey's chronic sinusitis, meant more antibiotics. Bacteria are most vulnerable to antibiotics at the moment they divide. Bacteria like strep divide rapidly, approximately every twenty minutes or so. *Borrelia burgdorferi*, however, divide much slower, and therefore dictate a need for longer courses of antibiotics.

It's imperative to make sure the antibiotic is present in the patient so that it's ready and waiting when the bacterial division occurs. While the generally accepted protocol for Lyme disease treatment is a monthlong dose of antibiotic, I usually prefer to treat for a minimum of six weeks, and frequently several months (especially in more persistent disease), to make sure I catch any stragglers.

Sometimes relatively short courses of antibiotics help initially, but the risk of Lyme disease recurring is high if you only try to eradicate the spirochete and not the round-body forms, the intracellular forms hiding in the cell, and the biofilms within which it can evade the immune system and antibiotics. Like an alien that takes over its host, Lyme will stop at nothing, morphing itself into whatever form it needs to remain undetected. To combat this, I prefer to treat with combination therapy, using antibiotics with differing mechanisms of action, some that can disrupt the cell wall of the spirochete in the extracellular space, some that can reach those that have become intracellular, others that can deal with those that have morphed into round-body forms, along with enzymes that can digest biofilm and allow the antibiotic to reach the bacteria hiding there, frequently administered twenty to thirty minutes prior to the antibiotics. I don't take any pleasure putting patients on multiple antibiotics simultaneously. I've spent much of my career treating people hurt by antibiotic side effects, like kids with recurrent ear infections and people with yeast infections and antibiotic-induced diarrhea. But sometimes, combination antibiotics may be the only way to truly eradicate the bacteria. For this reason, I also prescribed a high dose of good-quality probiotics to keep Stacey's gut healthy.

For the chronic sinusitis, Stacey started using a sinus nebulizer to deliver antibiotics and antifungals—with biofilm busters, because there are prominent biofilms in the sinus that can prevent the oral antibiotics from getting to the bacteria—straight to the site of the smoldering infection. I also prescribed a course of doxycycline and rifampin for the Bartonella. We used a PICC line (peripherally inserted central catheter; a small, flexible tube inserted into an arm vein that carries blood to the heart, used when a long course of IV medication is warranted) to deliver these intravenously to help mitigate her GI issues.

High-quality mercury-free and PCB-free fish oil has been shown to be an excellent anti-inflammatory, so Stacey started taking a dose of that every day, as well as vitamin D, curcumin, and NAC (N-acetyl cysteine), an amino acid that has shown some efficacy in treating stubborn OCD.

Now that we were tackling the underlying infection causing autoimmunity, we could get to work on calming Stacey's autoimmunity itself. She had seen incredible improvements with her first two IVIG treatments, albeit temporarily, but after each of the last four she had not seen as impressive results, and after the final one she had suffered so terribly from migraines she was hesitant to try IVIG again. I really wanted to see if we could re-create the positive changes she'd experienced after those initial two treatments. To protect against the migraines, I administered a nutritional cocktail of natural anti-inflammatories and magnesium, and followed her IVIG treatments by hydrating her with IV fluids and magnesium. Between the anti-virals that controlled her chronic viral issues and antibiotics to keep her underlying infections under control, the IVIG treatment again produced significantly positive results.

Results

Stacey noted improvement over the course of the next six to eight months. Her sugar cravings gradually decreased. She lost weight, and her mood stabilized by approximately 80 percent within two months, though she was still unable to attend school for more than a few hours at a time. She continued therapy for OCD, anxiety, separation anxiety, school phobia, and intrusive negative thoughts, but only now, after receiving the appropriate medical therapy, did she start to see improvement.

Thankfully, the separation anxiety almost completely disappeared. In time, her mother reported that she had become more social and even returned to gymnastics. She continued to lose the extra weight. Her rages and her tics disappeared. Rita couldn't believe it. She sat in my office, teary and emotional, marveling, "It's amazing. It's like my kid is back." Slowly, carefully, we tapered off the antipsychotic, Seroquel, and the SSRIs.

It wasn't always smooth going. The high dose of prednisone she took along with the IVIG treatment caused heightened emotionality that left her crying for hours on end, so we had to adjust. Rifampin is a very strong antibiotic and though we saw it made a huge difference in the Bartonella symptoms, Stacey developed some severe depressive side effects and we had to stop it. She did much better on Zithromax and Augmentin, though like with the rifampin, she had herx reactions after getting up to full strength Zithromax, including joint pain in her knees and wrists, as well as burning nerve pain in her shins, pain in her palms and heels, soaking night sweats, and occasional palpitations that persisted for one and a half weeks. Once she made it through this difficult herx, however, she showed a rapid positive mood shift.

While Stacey has made tremendous progress, it remains unclear whether she will ever fully recover from the effects of Lyme and Bartonella. In many ways, she's just like any other teenage girl. She was excited to get her driver's license. She's approaching the end of high school and thinking about college. But the uncertainty of never knowing if her moods will spiral out of control again can be stressful. The reality is that whether due to her genetic makeup or damage wrought by extended exposure to Lyme and Bartonella—or more likely, a combination of both—she will likely always have to be mindful of her environment. For example, a moldy bedroom ceiling in a summer house sent her tics into overdrive, and her separation anxiety worsened to the point where she wouldn't leave her chair, much less the house. It simply doesn't take much to throw her immune system off-kilter. Sinus nebulizer treatments with the antifungal medication itraconazole, along with biofilm busters, helped her get back to the improved state she was in before her vacation.

After hearing her story, one of Stacey's teachers, an Iraqi vet, assured her that the trauma they'd each suffered—him fighting in a war, her fighting through a debilitating medical condition—was not so dissimilar. She has a dream of creating a facility where ITABI kids could go for medical treatment and be completely psychologically supported. It would be a place where patients would never be disbelieved or blamed for their illness—no one would accuse them of exaggerating symptoms or suggest they're not trying hard enough to get better. This is something Stacey cares about deeply, as one of the most difficult parts of her recovery has been letting go of the pain inflicted by the judgment of others—not only doctors, but friends.

In the meantime, she takes her life one day at a time. One of the most complicated cases I've ever faced, Stacey is a testament to the importance of early detection and intervention, as well as a mother's undeterred resolve to find answers for her child.

A Welcome Development

An exciting thing happened in 2019, as I was in the middle of writing this chapter. The CDC announced that it would add a disclaimer to its stated criteria for Lyme disease, including on its 2017 Lyme disease case definition web page: "Surveillance case definitions are not intended to be used by healthcare providers for making a clinical diagnosis or determining how to meet an individual patient's health needs." I was relieved and elated. Finally, a public acknowledgment that the CDC criteria didn't have to be the *only* criteria for an official Lyme disease diagnosis. This gives me tremendous hope that rates of early detection of this tricky, complex disease will rise and save thousands of patients from undue suffering.

In addition, IDSA, the American Academy of Neurology (AAN), and American College of Rheumatology (ACR) released a draft of their new guidelines for prevention, diagnosis, and treatment of Lyme disease, opening it up for public comment. Although I don't agree with a number of the guidelines, including dosing of tick-bite prophylaxis and persistent disease, there were several positive changes, such as removing the strong recommendation that doctors not prescribe doxycycline for children under the age of eight. Previously, there was concern that doxycycline would mottle young children's teeth, and so the antibiotic of choice was amoxicillin. It turns out that concern was unwarranted, and now doctors are more likely to use doxycycline, which has the advantage of also being effective against coinfections.

As the debate over the best way to treat Lyme disease continues, I hope that both sides of the Lyme wars will eventually come together and discuss best practices with more openness and acceptance of their respective experience and expertise. It's the only way to ensure our patients will win back their health.

COULD YOUR CHILD HAVE LYME DISEASE AND
ONE OF ITS COMMON COINFECTIONS?

Where does your child live?

Where has your child vacationed?

Does your child play outdoor sports?

Does your child enjoy running around in the grass, or walking, hiking, or playing in the woods?

Have you ever found a tick on your child?

Have you or your child ever noticed an unexplained rash on the body?

Have you ever noticed your child with an unexplained swollen joint, or has your child ever complained of difficulty walking?

Does your child complain of significant fatigue?

Has your child ever had a summer "flu" or flu-like illness?

Do you have pets that walk or play outdoors?

Did your child have frequent ear, sinus, or other infections that required frequent rounds of antibiotics?

Has your child taken a fairly recent hike in the woods, or taken a trip to the shore or mountains, or had a stint at summer camp?

Has there been a tick bite or a red circular-type rash?

Have you noticed that your child has developed violaceous (reddish-purplish) stretch marks that do not correlate with recent weight gain or loss, and are in atypical places for stretch marks, such as the middle of the back or behind the knees?

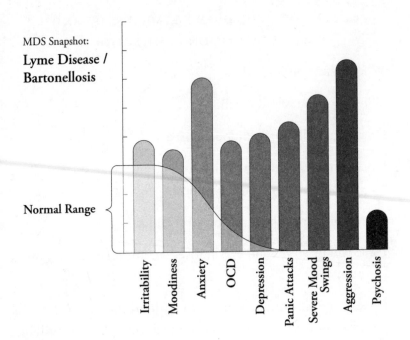

MDS Snapshot:
**Lyme Disease /
Bartonellosis**

Normal Range

Irritability
Moodiness
Anxiety
OCD
Depression
Panic Attacks
Severe Mood
Swings
Aggression
Psychosis

CHAPTER 6

Stopped Cold

Hashimoto's and Other Thyroid Disorders

By the time Julie Gérard and her family stepped out of the rental car in my Red Hook office parking lot, the French native had been trying to get answers from pediatricians in Germany, where she lived with her husband and two children, for five years. Why was her daughter, Chloé, so tiny when Julie was average height and her husband on the tall side? Why did she have what Julie thought of as "mood problems," high one minute, then unnaturally low the next, for extended periods of time? Why was she either hyperactive or too tired to play, and rarely anything in between? Why did she suffer so much from intestinal problems, switching between diarrhea and constipation from month to month? Why was the girl so exhausted after a few hours of kindergarten that she couldn't even eat? Chloé's fatigue and irritability made it hard for her to make friends; Julie could see that even the child's teachers had difficulty warming up to her. Chloé was tested repeatedly for all the obvious culprits—celiac disease, low blood sugar, food sensitivities, nutritional deficiencies—and nothing revealed itself. The pediatrician did eventually confer an

ADHD diagnosis, but the medicines typically prescribed to treat it didn't seem to help Chloé much.

Julie kept asking whether there could be a link between her daughter's physical, cognitive, and emotional issues, yet every doctor she spoke to assured her they were not connected. Julie wouldn't accept that her daughter could not be helped. Despite the fact that Chloé hadn't tested positive for gluten intolerance, Julie put her on a gluten-free diet to see if it might help alleviate her GI symptoms. It did. Chloé was much more regular and comfortable, and her moods seemed to even out a little. But the difference wasn't dramatic, and Chloé's discomfort and crankiness didn't disappear. When Julie brought her observations to Chloé's doctor, he was appalled that Julie would put her daughter on a restricted diet and demanded she reverse course. Julie suspected that every physician they met thought *she* was her daughter's biggest problem, and soon she stopped asking questions, pressing for answers, or even answering doctors' questions with complete honesty. Her confidence was so shaken, she began to wonder if maybe they were right.

Around the time Chloé turned five, her two-year-old baby brother was diagnosed with asthma. Julie wasn't sharing her concerns with doctors anymore, but that didn't mean she'd stopped doing research. One day she Googled the words "ADHD" and "Asthma" side by side, and up popped a book that discussed the common root causes of four of the fastest-rising childhood disorders: autism, ADHD, allergies, and asthma, or the 4-A disorders. It was my book, *Healing the New Childhood Epidemics: Autism, ADHD, Asthma, and Allergies.* Maybe, Julie thought, the author of the book would at least *consider* the possibility that all of Chloé's symptoms were related. She had to find out.

That's how about six months later, she and her husband found themselves sitting in my office, trying to shake off their jet lag, while Chloé and her baby brother played with toys on the floor. Julie already knew this was going to be a different type of doctor's visit— the preconsultation questionnaire alone had taken over two hours to complete. After a long interview covering everything from Julie's pregnancy all the way to the current day, I started examining Chloé. As I conducted the exam, Julie noted that I thought out loud, reciting all the symptoms Julie and her husband had listed. Julie was surprised to see me particularly focused on the fatigue, ADHD, and mood symptoms. Then Julie heard me utter words no other doctor had ever said: "Let's take a look at her thyroid."

How the Thyroid Works

The thyroid—a little bow-tie-shaped gland wrapped around the trachea at the base of the neck—has an oversized influence for such a small organ. It serves as the control center for the metabolism of every single cell in the body. That is, it's responsible for managing the entire range of biochemical processes in the body that turn food into energy. With such all-encompassing reach, a thyroid disorder can quietly restrict a child's physical, digestive, and cognitive growth and development. And because in many cases symptoms build up extremely slowly, it can be easy for a thyroid condition, especially low thyroid, to be overlooked.

The thyroid does its job by using the iodine present in food and combining it with an amino acid called tyrosine to convert it into two thyroid hormones: triiodothyronine (T3), so named because it consists of three iodine atoms per molecule, and thyroxine (T4), which, go figure, consists of four iodine atoms per molecule. These hormones, attached to carrier proteins, travel through the body via the bloodstream and enable cells to convert oxygen and calories into the energy they need to function, thus powering or regulating the digestive tract, heart function, brain development, growth, muscle contraction, body temperature, and cell regeneration. In actuality, the active components are the free unbound parts of these thyroid hormones, referred to as free T4 and free T3.

T3 is the active hormone that keeps your body functions and growth working normally, whereas T4 has to be converted before it's of any use. Through a process called deiodination, organs such as the brain, liver, and kidneys use a selenium-containing enzyme to clip off one of the T4 iodine atoms and create their own usable T3. What this means, in sum, is that selenium and iodine are critical to proper thyroid function. The T3 then binds to receptors on or in cells, activating cell signaling pathways to send a message so the cell can carry out its particular metabolic functions.

Normally, the hypothalamus, the pituitary gland, and the thyroid are locked in a precisely calibrated feedback loop to ensure the body's hormone levels are exactly where they need to be for optimal organ function. The hypothalamus, located at the base of the brain stem, releases thyrotropin-releasing hormone (TRH), which tells the pituitary gland attached directly beneath it to release thyroid-stimulating hormone (TSH), which then tells the thyroid to create more T3 and T4. When the thyroid is producing sufficient amounts of T3 and T4, it sends a message back to the pituitary directly to decrease its production of TSH, as well as to the hypothalamus to stop producing so much TRH. But sometimes, due to environmental impacts, disease, genetic predispositions, or a combination of all three, communication can break down and this feedback loop can be disrupted, resulting in the thyroid producing too much or too little T3 and T4.

How often does this happen? According to the American Thyroid Association, about 20 million Americans suffer from a form of thyroid disease, but the symptoms creep up so slowly, and are so broad and easily attributable to so many other conditions, that about 60 percent of those people are completely unaware they have it. Thanks to improvements in thyroid testing sensitivity, many people who would have once fallen through the cracks are now getting diagnosed. Without these sophisticated tests, thyroid disease is hard to pinpoint. It's another Great Masquerader, second only to diabetes as one of the most common endocrine disorders.

What to Look For

The physical symptoms of hypothyroidism include fatigue, sensitivity to cold, dry skin, hyperkeratosis (small bumps on or scaling of the skin), depression, weight gain, hair loss, coarse, dry hair, constipation, and heavier than usual or irregular menstrual periods. You can see why it's easy for this diagnosis to fall through the cracks. After all, who doesn't complain of fatigue nowadays? One unmistakable sign, however, is an enlarged thyroid, also known as a goiter. Until the early twentieth century, when treatments became readily available, goiters were often left untreated and grew to massive sizes. It's easy to look at examples, since goiters have been depicted in a range of art forms across cultures and eras, some as far back as 500 BCE, with a number in evidence in Renaissance works, including some drawings by Leonardo da Vinci. Today, goiters are uncommon in the U.S., and except for a few very noticeably enlarged thyroid glands, the ones I generally see are just a little plump, noticeable only when I palpate the gland around the base of the neck.

Sometimes, however, in addition to these physical signs, patients experience cognitive and neuropsychological symptoms such as brain fog, poor memory, anxiety, depression, excessive sleepiness and fatigue, and low concentration. Most of these overlap with features of mood disorders. Adult patients have been known to experience auditory and visual hallucinations and suffer delusions. In one case study, after burning himself with acid in an attempt to be "disconnected from the universe," a patient who had been struggling with manic and depressive episodes since adolescence was found to be severely hypothyroid. To his recollection, no one had ever checked his thyroid levels. In fact, 1 to 4 percent of patients with mood disorders like bipolar disorder and depression have classic hypothyroidism. While this particular patient was treated with a combination of an antipsychotic and the synthetic hormone levothyroxine, in some cases cognitive and psychiatric symptoms have been reversed with thyroid hormone replacement alone.

Where's It Coming From?

Women are five to eight times more likely than men to develop a thyroid disorder, especially during life phases when their hormones are dramatically in flux, such as pregnancy, postpartum, and menopause. The risk generally rises with age, and yet two to three of every hundred children will be diagnosed with hypothyroidism, most often due to Hashimoto's, the autoimmune disease that is the primary cause of the condition (researchers believe women are more likely than men to be affected because they are also more susceptible to autoimmune disease).

Hashimoto's thyroiditis is extremely common—of the 20 million people in this country with thyroid disorders, 14 million have Hashimoto's. What could be causing it? Stress is a possibility. It's well established that stress can affect the immune system. It can suppress it, or at other times trigger it to become overactive (i.e., dysregulated), and lay the groundwork for autoimmune disorders, a known common cause of thyroid disorders. And there's no question everyone, kids included, is living with far more daily stress than previous generations. Another possible cause is Lyme disease, a known trigger for immune dysregulation. I treat a lot of Lyme patients, so I see a tremendous number of children with autoimmune thyroid disease, specifically Hashimoto's. Stacey was one. Otis, a little boy with a round face framed by Harry Potter–style glasses, was another.

The Wild Child

Otis came to me to be evaluated for possible nutritional, hormonal, and metabolic imbalances while he was simultaneously being treated for Lyme by another doctor closer to home. I always looked forward to his visits because he was so cute and sweet. He was so adorable it was almost impossible to believe he was capable of the monstrous behavior his mother described. This child, who regularly sat in the cafeteria with a classmate with cerebral palsy whom other kids rejected and teased because he couldn't keep food from spilling out of his mouth while he ate, was the same child who one day pulled a knife on his mother. Months of oral antibiotics seemed to calm his symptoms, but each time his doctor would take him off the medication, citing concerns about side effects, the boy's rages and aggression would return. The family lived in terror for themselves and for their child. Eventually they turned over Otis's Lyme treatment to me.

He was eight years old the first time I met him, and a thyroid test showed that not only were his T3 levels low, he had elevated levels of thyroid peroxidase (TPO) antibodies. The presence of antibodies suggests an autoimmune disorder—in this case, possibly as a response to the Lyme disease, Otis's body had mistakenly started attacking his thyroid. He was thin, which is unusual in someone with low thyroid since it slows down the body's metabolism, often causing weight gain, but he had many of the other physical symptoms that correlate with Hashimoto's: constipation with very large stools; unusually short stature for his age; eczema; keratosis pilaris, also called "chicken skin"—small, dry, usually flesh-colored, occasionally slightly reddish bumps that frequently appear on the outside of the upper arm; high cholesterol levels at 181 when they should have been no more than 170. Finally, he had several cognitive symptoms that can accompany Hashimoto's, such as problems with memory, attention to and retention of learning, decreased energy, and fatigue. His behavior was impulsive and irritable.

On their own, these cognitive and behavioral symptoms would not necessarily be connected to Hashimoto's—the boy did have Lyme disease, after all—and the symptoms of thyroid disorder are notoriously vague and overlap with many other conditions. But Otis's particular physical symptoms raised the possibility that a thyroid disorder could be causing his cognitive issues, or at least contributing to or exacerbating issues being caused by the Lyme disease.

What's Normal?

To review, Otis's T3 and T4 levels were in the low-normal range and his TPO antibody levels were elevated. Interestingly, his TSH was normal, which means he didn't present as a classic picture of hypothyroidism. As with everything, thyroid disorder is on a spectrum.

We have assigned numbers to "normal" TSH and T4 levels, yet people can be low-normal, borderline, or high-normal and still be perfectly functional. The term for this is "subclinical," yet I disagree with this nomenclature, especially in patients with multiple vague symptoms. When the labs show evidence of thyroid dysfunction but the patient has no symptoms, that may be termed subclinical. But if the symptoms are present, it's not subclinical, it's sublaboratory. The symptoms are surely there, they're just not being recognized as thyroid disorder symptoms. Then again, some people with lower than average or higher than average hormone levels can show very significant symptoms. The degree of thyroid dysfunction noted on labs does not always correlate with the severity of symptoms.

So what's considered "normal?" Traditionally, a lab provides a reference range for all the different thyroid related hormones—TSH, T4, T3, and free T3 and free T4 (the active unbound components). These ranges vary from lab to lab, which is important to know when you evaluate thyroid hormone results. Because there's a range, a very low thyroid hormone level or a significantly elevated TSH could still be considered normal. Typically, doctors will only treat a patient for hypothyroidism if the TSH is definitively above the upper limit of the range and the T4 or free T4 is below the lower limit of the range. That's standard, especially if the patient exhibits symptoms. But I do things differently.

First of all, in addition to checking free T4 levels, I check free T3 levels. If I see normal or low-normal free T3 or free T4 levels with elevated or high-normal TSH, I can't dismiss it, especially if I see corresponding symptoms. A child with dry skin, constipation, and fatigue is having hypothyroid symptoms, even if her labs say she shouldn't be. Hashimoto's symptoms can precede abnormal thyroid blood tests by many months. Regardless, for whatever reason, this individual's immune kettle is such that, for this body, the thyroid levels are insufficient.

My goal is to get the contents of a person's immune kettle as low as it can go so the body can be as healthy as it can be, with increased resilience, so it doesn't have to fight on as many fronts. For this reason, it seemed prudent to put Otis on a very low dose of thyroid hormone along with the antibiotic regimen that would help eliminate the infection caused by the Lyme.

Otis's symptoms improved. We eased him off the antibiotics until his only regular medication was a nightly low dose of Depakote, prescribed by his psychiatrist to help control the migraines he sometimes suffered, and thyroid hormone replacement. Initially he was started on T3 (L-thyronine), as repeat testing showed decreased free T3 in spite of low-normal free T4 and TSH. As is required with Hashimoto's thyroiditis and hypothyroidism, I regularly followed his thyroid function, and when the free T4 also eventually came back low (still with normal TSH), we added T4 (Synthroid). As we gradually increased his dose, Otis gained four pounds and grew a few inches. His mother reported that he was back to playing outside with his friends, and while he still had large bowel movements, they became more regular and he was finally less irritable. I stopped seeing him regularly once he turned twelve. He progressed beautifully through his middle and early high school years, thanks to the support of his parents and a well-structured special education program. He was on the honor roll, had plenty of friends, and volunteered at the local senior center.

Then things started going south again.

One major issue when treating older children is cooperation and compliance. When children are young, their parents can control their diets and ensure they take all of their medications and supplements. Once they get older and start to become more independent—or come down with a case of teenagitis—they frequently become less consistent or stop taking their medicines altogether. They don't enjoy bringing their own foods to parties or hearing jokes about their lunch in the school cafeteria. Sometimes they simply get tired of having to remember to take pills all the time. Or maybe they just want to ignore any reminders that they're different from their peers. For whatever reason, eventually Otis stopped taking his medications.

When his mother called to say that she was seeing a resurgence of some of Otis's aggressive behaviors, he was about seventeen years old. He'd shown no signs of aggression for several years and had accepted rules and limits his whole life; to start rebelling, dropping old friends, smoking pot, and defying his parents at this late stage, when he was almost done with high school, seemed strange. He was angry, moody, and irritable. Even his new friends were commenting about his change in temperament. Within two weeks of that call, Otis stopped going to school, and when he did go, he was disruptive. One day the school ordered him to detention, and his mom started receiving suicidal texts from him. When she called the school to find out what was going on, administrators couldn't find him. They put the building on lockdown, but he'd already escaped. Otis's mom told the school to call the police. Otis didn't show back up until later that evening, when he came in through the back door looking haggard, saying he just wanted to go lie down. While his father calmly talked with him in the kitchen, Otis's mother called the police, who told her to leave the front door unlocked so they could come in without alerting Otis. Incredibly, when he saw them, Otis didn't run, and the police were able to escort him to an ambulance waiting outside, which took him to the local children's hospital. It was only after several days there that he was willing to see me again.

His test results showed renewed inflammation, with a spike in his antithyroid antibodies. I put him back on T3 (slow-release liothyronine) and T4 (Synthroid), and restarted Depakote at night. I spent a great deal of time with him discussing the relationship between his noncompliance and the recurrence of his symptoms. Teens and young adults want to forget they were ever sick, but rebellion can have serious repercussions for kids with medical conditions. It's only when their symptoms start to reappear that they remember how awful they used to feel, and it's only when they start feeling bad enough that they come back in. Thankfully, as we had already established a good rapport, he took our talk to heart and followed my treatment recommendations. Slowly his symptoms started to ebb.

Next time I saw him, he was twenty years old, and without even testing him I was sure I'd find low thyroid. His moods had evened out and he had started community college, but he complained of low energy, insomnia, fatigue, constipation, dry skin, and a scalp so dry he had dandruff. He said his mind was always racing, making it hard for him to focus. He initially denied depression, but the more I spoke with him, the more he seemed to be emotionally and psychologically weighed down. Sure enough, upon testing we discovered that even though his thyroid function seemed fine on paper, his thyroid antibodies were high. He also finally admitted that he'd stopped taking his supplements and his thyroid hormone replacement. Concluding that Otis was experiencing all the symptoms of low thyroid, I put him on a dose of natural thyroid (Nature-Throid), and within a couple of weeks, his rage deescalated.

Due to the complexity of his case and the multitude of his symptoms and contributing underlying factors, it's impossible to say that Otis's Hashimoto's thyroiditis and thyroid dysfunction directly caused each specific symptom he experienced. However, there is no doubt that each time we treated his thyroid, he was better able to manage all his other ailments. The gluten-free diet helped his anxiety and irritability; antibiotics enabled major improvements with the mood dysregulation, violence, and anger; but the thyroid, in addition to contributing to those overall improvements, clearly helped increase his growth, his energy, his bowel regularity, and his concentration.

Environmental Toxins

An argument can be made that the chemicals in our environment are responsible for the rising levels of thyroid disorders around the world. The thyroid gland seems to be particularly vulnerable to the effects of elevated levels of toxins in air, water, and food supplies, though there is still much research to be done to ascertain how and to what degree certain chemicals affect thyroid function. Some studies have shown that individual chemicals have no negative effects except when mixed with other substances, which is the more likely scenario outside the lab. For example:

- Some chemicals like polychlorinated biphenyls (PCBs, banned in the late 1970s but still present in our environment and many food sources) seem to be able to suppress thyroid hormone levels without affecting TSH, indicating that they're influencing hormone action somewhere other than the pituitary or thyroid gland. Others, like dioxins—by-products of certain manufacturing processes and the production of herbicides and pesticides that are mostly found in meat, dairy products, fish, and shellfish—have been shown to increase TSH levels in humans.
- Bisphenol A (BPA, used in many plastics, including the linings of some metal food cans) can bind to thyroid receptors, thus keeping T3 hormones from transmitting their messages to cells. That's how a patient can have labs that show normal levels of T3 yet experience all the physical symptoms consistent with low thyroid.
- Perchlorate, used in making rocket fuel, fireworks, and other explosives, and known to bind to and block iodine receptors, has been found in public drinking water and contaminated foods.
- Fluoride has been known to "compete" for receptors on the thyroid gland, making it impossible for the thyroid to 1) respond to the pituitary's TSH, thus reducing the creation of hormones, 2) block TSH production from the pituitary gland, and 3) interfere with the deiodination process that allows organs to remove an iodine atom from T4 to create usable T3. If fact, if you remember your periodic table from high school, fluorine, chlorine, and bromine all line up in a column on the right side. What's the element pictured directly below them? Iodine. These four elements, called halogens, are clustered together because they share a similar molecular structure. Iodine's position below the other three also means it has the

heaviest atomic weight. Between the molecular similarities and its heavy weight, it is easily "displaced" by the other three halogens. The thyroid often can't quite tell the difference between them. When our thyroid mistakenly stores the fluoride from our water and toothpaste, chlorine from our swimming pools and showers, or bromine found in flame retardants, plastics (it's the source of "new car smell"), and some packaged baked goods, it does so because it thinks it's absorbing iodine. Unfortunately, those other elements can block the creation of T3 and T4 as well as their ability to bind to receptors.

Food Sensitivities

Finally, food sensitivities such as gluten or dairy (which we'll discuss at length in chapter 8) can also trigger autoimmunity, which, as you'll recall, is a common cause of thyroid disorders. Increased intestinal permeability can cause autoimmune disease, and I've seen a pattern of patients with leaky gut also present with Hashimoto's disease. In one study, almost 30 percent of patients with non-celiac gluten sensitivity also developed autoimmune disease.

On the Flip Side: Hyperthyroidism

While hypothyroidism is the most common thyroid disorder, 15 percent of childhood thyroid disorders are due to hyperthyroidism, or excessive thyroid hormone production, mostly caused by Graves' disease, an autoimmune disorder. Physical symptoms can include warm, moist skin, muscle weakness, hand tremors, a goiter, a bruit (the sound the additional blood flow makes over the enlarged thyroid), overactive reflexes, sensitivity to heat, depression, diarrhea, and lighter than usual or even missed menstrual periods in girls. Puberty can be delayed. Because other symptoms frequently mimic those of anxiety or ADHD—rapid irregular heart rate, weight loss, nervous energy, insomnia—it can take awhile for a patient to get an accurate diagnosis. Hyperthyroidism is sometimes confused with respiratory disease or cardiac arrhythmia. Though mania has been reported in cases of hypothyroidism, it's far more common when thyroid hormones are elevated.

Certainly in adolescents with anxiety, hyperactivity, insomnia, and bipolar disorder, I consider and test for hyperthyroidism, but I must say the mainstay of thyroid dysfunction I see in my practice is hypothyroidism.

Treating Thyroid Disorder

Fortunately, for all the havoc that little bow-tie-shaped gland can wreak on the body, treating thyroid disorder is rarely complicated.

Thyroid Replacement

To reverse hypothyroidism caused by Hashimoto's, most doctors prescribe an inexpensive synthetic hormone, levothyroxine, often sold under the brand names Synthroid or Levothroid, to supplement the amount of T4 your thyroid is unable to produce. Once the levels are normalized, the feedback allows the hypothalamus and pituitary to stop overproducing TRH and TSH.

In many cases, especially in those with sublaboratory hypothyroidism, I actually prefer to use a natural thyroid replacement, usually produced from desiccated pig thyroid, to also enhance the amount of T3. It's suspected that the higher success rate for natural thyroid isn't because it's natural, per se, but because natural thyroid contains T3, whereas synthetic T4 does not, in addition to building blocks like selenium and tyrosine. In fact, recently more and more doctors who rely on synthetic T4 have started adding a dose of synthetic T3—usually in the form of liothyronine—to their daily treatment. Notably, there have been cases in which the mood of a patient with long-standing depression was suddenly lifted just by adding T3 to a typical T4 treatment.

Nutritional Supplements

While additional pharmaceutical hormone support is generally necessary, when it's a borderline case it is sometimes possible to treat low thyroid with nutritional supplements to support healthy thyroid function. As we established earlier, iodine is one of the key components of proper thyroid function. Fortunately, it's also one of the easiest and least expensive treatments for low thyroid, found in iodized salt on supermarket shelves across the United States. Iodine deficiency is generally thought to be rare in this country, though it remains common worldwide. However, I don't think it's rare. I generally check iodine levels in my new patients and frequently find borderline to frank iodine deficiencies.

We also established that another essential nutrient for healthy thyroid function is selenium, the active ingredient of the deiodinase enzyme that cleaves one of the four iodine atoms of T4 to make it into usable T3. If you're low in selenium, which many people are, you may not have an active enough deiodinase enzyme, and therefore you may not make enough T3, even though your thyroid is producing healthy levels of T4.

I also use a combination nutritional formula that in addition to iodine and selenium includes zinc, vitamin A, vitamin D, and vitamin E. Antioxidants such as vitamin E, along with selenium, can help support the thyroid by reducing oxidative stress. (See appendix for more details.)

Gluten-Free Diet

We'll talk more about the benefits of a gluten-free diet in chapter 8, but given the correlation between Hashimoto's patients and gluten sensitivity, it makes sense to consider this could be a helpful treatment for low thyroid. Researchers have been working on studies that support the anecdotal evidence reported by many non-celiac Hashimoto's patients who claim to have seen improvements with gluten-free diets, and a more recent study confirms this.

Today many endocrinologists are discussing the possible benefits of gluten-free diets with their Hashimoto's patients. If antibodies are attacking your child's thyroid and you want to figure out what could be driving it, it's worth investigating the potential connection to gluten. It's easy to reverse course with no harm done if symptoms and/or antithyroid antibodies don't abate, as long as it's done in the context of an overall healthy diet, ideally with the help of a nutritionist.

In the end, it was a combination of these treatments that worked for Chloé. When the blood test results came in, they showed that her thyroid levels were borderline low-normal. She had tested negative for thyroid antibodies, so we knew we weren't dealing with Hashimoto's. I started by prescribing an extremely low dose of T4 and asked them to cut dairy and sugar out of Chloé's diet in addition to gluten.

After several months, Julie reported that Chloé's digestive issues had improved, but nothing else had changed much. So we tried a dose of T3. Not long after, I got a call from a stunned Julie. Chloé was different. Before, she spoke almost haltingly, couldn't focus, and seemed almost sluggish. Julie and her husband frequently had to repeat themselves and explain things to her very slowly. All of a sudden, Chloé had brightened up. Everything had quickened, from the way she spoke to the way she walked and drew. She was capable of much more complex thought processes and seemed to understand what was said to her without any trouble. She could concentrate. And her mood was up in a way her parents couldn't remember ever witnessing. It was like lifting a blanket and discovering a whole new child underneath.

Julie knew the transformation wasn't just a figment in the minds of two parents desperate to see their daughter improve. They had recently returned from a vacation in Greece with friends who had a daughter the same age as Chloé. Normally, the other little girl won every game the children played, and Chloé was always trailing behind her friend. On this trip, Chloé could beat her friend. She took initiatives and she could keep up. It was so dramatic, Julie's friends had commented on what a changed child Chloé was.

Unfortunately, the improvement was short lived. Soon after the family returned from vacation, Chloé started to regress and display more aggressive behavior. Thinking it must be an adverse reaction to the T3, Julie stopped giving her the medication. But as it turned out, the sudden change wasn't a reaction to the T3, it was a reaction to chlorine. While in Greece, Chloé had gone swimming almost every day in a chlorine-treated pool. Eventually Julie started to notice that every time Chloé went swimming in a pool treated with chlorine, her moods soured. But for two to three years, Julie didn't make that connection, and so Chloé didn't take the T3. Around the time Chloé turned eight, things got worse. She had started taking piano lessons, but she was too tired to play. She was incapable of keeping up at school and struggled to finish homework. Getting her out of bed in the morning was like a tug-of-war.

That's when I heard from Julie again. I told her to get a new set of blood tests, including of Chloé's thyroid. This time, no one, not even their pediatrician, argued with the results. Chloé was clearly hypothyroid. Her T4 levels had dipped well below normal. At first we started her on T4, same as before, which helped, but not as much as we would have liked. Julie was reluctant to try the T3 again because she hadn't yet figured out that it was the chlorine, not the hormone, that had sent Chloé's moods into a downward spiral, but she finally agreed to try a combination of T4 and T3. Within about a week, there she was again—the sharp, quick Chloé too few people had ever had the pleasure to know.

In the years since we switched Chloé to Nature-Throid, a natural thyroid replacement made from desiccated pig thyroid that contains all the elements found in any healthy thyroid (including T4 and T3), she's made a remarkable turnaround. Julie also observed continued improvement when we added vitamins B-6, P-5-P, methylfolate, and a nasal spray form of vitamin B-12 to Chloé's regimen in addition to her gluten-, dairy-, sugar-free diet.

Now at twelve years old, Chloé does well in school and takes great joy in her dance classes—jazz and ballet—and a new sport, bouldering. It's hard for Julie to relax and accept that her daughter's improvement could be the new normal forever. One day perhaps she'll be able to take Chloé's smile for granted, but for now Julie remains grateful for every good day her daughter has.

Look for Patterns

It's exceedingly rare for an illness or disorder to express itself without any warning. We may not feel symptoms or recognize them for what they are in the moment, but they're usually there. We don't walk from one room where we are disease-free into another one where we are suddenly sick. We're always living on a spectrum of health, our bodies in a constant state of flux as billions of biological messages and connections get transmitted or missed on any given day.

With thyroid disorder as with almost any other condition, symptoms creep up on you. They can be months in coming—slow, steady progressions that make it hard to connect the dots. But unlike Lyme disease or ITABI, simple awareness of the symptoms of thyroid dysfunction can put what seems like teen depression or anxiety in a whole new light. That's true for parents and for doctors. With more than three decades in the field, people ask me about retirement more and more these days. I can't imagine why I would want to retire—I love what I do. But in addition, I believe I'm a better doctor because of my thirty-five-plus years of experience. That's thirty-five years of pattern recognition that I can apply to more quickly help people.

Look for the patterns. If your child suffers from depression or cognitive issues, look for any physical symptoms—particularly dry skin, constipation, or fatigue—that could point to a dysfunctional thyroid, even if lab work doesn't show irregular results or reveals only borderline normal levels. What's normal for most people may not be sufficient for your child. Tests rarely tell a whole story. That said, identifying thyroid disorder isn't necessarily the key to everything. It can be an important piece of a diagnostic puzzle, though, and can prevent many kids from getting fully better if not addressed.

COULD YOUR CHILD HAVE A THYROID DISORDER?

Does your child frequently complain about feeling tired?

Have you noticed your child gain an inexplicable amount of weight?

Have you noticed your child lose an inexplicable amount of weight?

Is your child's hair coarse?

Do you notice dry skin, particularly small dry bumps on the outside of the arms or tops of the thighs? How about dryness and scaling of the lower legs, or hyperkeratosis (thick scaly buildup) of the soles of the feet?

Does your child complain of being cold or hot when everyone else around is comfortable?

Does your child complain of (or have, and not complain of) cold hands and/or feet?

Is your child frequently constipated?

Are your child's cholesterol levels elevated beyond the normal range for their age?

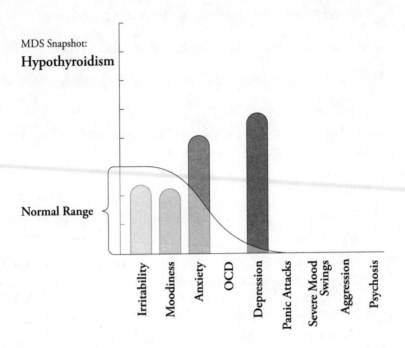

MDS Snapshot:
Hypothyroidism

Normal Range

Irritability　Moodiness　Anxiety　OCD　Depression　Panic Attacks　Severe Mood Swings　Aggression　Psychosis

CHAPTER 7

Dead on Their Feet

*Adrenal Dysfunction
and Low Blood Sugar*

I met the Newmans on a beautiful spring day in 2016. It's common for families to see me together when they're coming from out of state, and this one was seeking help for their severely autistic son, Will. At sixteen years old, the tall, dark-haired boy was in constant motion with self-stimulatory behaviors such as rocking and spinning his arms. Though nonverbal, he emitted explosive screeches. His entire life, he'd been prone to violent meltdowns that would turn the house upside down, the kind that would evoke looks of shock and disapproval from strangers on the rare occasions the family tried to do something together outside the house. As Will grew into a man, it was becoming harder and harder for his parents to calm him. For years they'd walked on eggshells so as not to do anything that might trigger their sensory-overwhelmed boy, reaching out to every doctor and specialist they could find who might be able to offer a new thread of hope. They'd heard that I had helped other autistic children break through their isolation and explosiveness. Was there anything, anything at all, I could do to help their son?

As our first visit together neared its conclusion, I looked into these parents' worried faces and let them know that I certainly hoped I could make a difference, and I would do everything I could. However, it was too early to tell them that it wasn't only Will who concerned me, but also their other son, Charles, the one quietly sitting across from them on my office couch.

I often refer to the families of my patients as the walking wounded. Their children may be the ones afflicted with a debilitating disorder, but the suffering of parents and other family members who love these kids can be almost as great. I don't think I've ever met a mother or father who wouldn't willingly remove their child's illness or disorder and place it upon themselves if it would ease their child's pain and fear. Nothing takes precedence over trying to help their sick child. I've known parents who've quit their jobs to devote themselves full time to their child's care, or drastically downsized their lives to afford costly medical treatments. Unfortunately, sometimes this focus on caring for the sick child in the family can rob other children of the attention and care they need to thrive.

I've seen this play out many times over the years—siblings who get short shrift because so much of their parents' time and energy is invested in their brother or sister. Of course this emotional neglect is never intentional, and according to some studies, many parents genuinely believe their other children are coping well—not because they're blind to their children's needs, but because their healthy children learn to shape their needs and identities in a way that works for the family. When healthy children are involved in a "cohesive family unit" and allowed to participate in appropriate ways with their sick sibling's care, it can bring families closer together. But when siblings are shut out of that care and left in the dark, they tend to meet silence with silence. These children learn that the best way to help their families is to make the fewest demands possible, which means they don't draw the attention to themselves they normally would if a problem was brewing.

During that first visit with Will, his parents and I covered every detail of his symptoms, behaviors, and the ins and outs of their daily lives. Indirectly, I learned that twelve-year-old Charles, who was neurotypical, had lived his whole life in the shadow of his autistic brother. Will's outbursts and meltdowns were upsetting to adults; I can only imagine how frightening they would be to a young child. It had to be stressful, living in a house where you never knew how long peace and quiet would last. Where you could count on your parent leaving you midconversation to deal with your brother. Where you never came first. The entire family was in constant upheaval. In my experience, most children's way of dealing with chronic instability and stress is to either act out or retreat.

Charles didn't look up from his iPhone while I interviewed his parents. Once they'd given me the information I needed, I turned my questions to the small-statured, dirty-blond boy on my couch. How was he feeling? How was school? Did he like hanging out with his friends? All of his answers were exactly what you'd expect to hear. "Fine." "Good." "Yes."

I've thought hard about how to explain that despite appearances, I was absolutely sure that all was not fine or good with Charles. Perhaps it was the meek smile that didn't reach his eyes. It wasn't just that his body and voice lacked energy, it was that he was completely devoid of spirit. He'd disappeared so far into himself that it was impossible to get a read on his personality. I glanced over to his parents to gauge their expressions. They were watching him benignly. Often a mom or dad will pipe up with an observation like, "She's a great soccer player!" or "He's becoming an incredible cook," something that proves to their child (and the doctor) that they're paying attention. Neither spoke, and we all turned back to Will.

But I didn't forget Charles. I saw him every six to eight weeks when the family would return for follow-up visits to adjust Will's treatment as necessary. Within a few months of starting his treatment plan—including dietary modifications, nutritional supplementation, and mood-stabilizing medications—we began to see glimmers of improvement in Will. More eye contact. A few more words. Fewer explosive episodes. I'd turn my attention to Charles as I got my update from the family, and he always appeared the same—fine. His answers to my questions about school, friends, and interests seemed scripted, like he was performing an act he'd perfected over many years. It was so perfect it was almost overdone. Where was the normal grumpiness about homework? Why was there no spark when he talked about sports or video games? His emotional range had flatlined. Yet his parents, so consumed with worry for their older son, could not see that their younger boy was hurting, too.

As in medical cases where I have to stem the immediate danger before I can turn my attention to treating less dire but equally important dysfunctions, it wasn't until I was confident that Will's progress was gaining traction and that the family's home life had steadied, that I gently raised my concerns about Charles. These were good parents, and I didn't want them to feel judged or threatened. I simply hoped to help them see that they'd gotten caught in a hellish cycle and needed help to find their way out of it. I hoped that with the right approach they'd see for themselves that something wasn't right.

The way to turn things around for Charles, I believed, was to remind his parents that their younger son had also gotten caught up in this awful cycle, and that it's not uncommon for parents to underestimate how much stress and anxiety a child is experiencing. In this case, I explained, I suspected that after years of trying to help Charles cope with the frightening, destabilizing situation at home, Charles's stress managers—the adrenal glands—had lost their ability to do their job.

The Adrenals: What Are They?

As fellow members of the endocrine system that work in tandem with the thyroid, the adrenals function as a major communications hub. These glands, located just above the kidneys, talk to the pituitary gland and the hypothalamus in the brain. Together, these organs form the hypothalamic-pituitary-adrenal axis (HPA axis), and they coordinate to send out hormones that manage metabolism, suppress inflammation, support the body's immune system, and maintain proper sodium balance. In another fascinating example of the body's interconnectivity, the HPA axis, which allows the nervous system to interact with the endocrine system, is also responsible for the fight-or-flight response you experience when your brain perceives a threat.

For example, when you are startled or frightened by a life-threatening situation, say by a car that speeds too close when you're walking, your heart rate jumps, your breathing speeds up, your vision sharpens, your muscles tense, and you start to sweat. That happens because in an instant, your hypothalamus interprets the speeding car as a threat and transmits that information to the pituitary gland, which then tells the adrenals to release a surge of hormones, including adrenaline. That chain reaction preps your body to save itself by flooding you with the speed and energy you need to jump out of the way of the speeding car. In addition to releasing adrenaline in response to a perceived stress or threat, the adrenals release a steroid hormone called cortisol. Under normal circumstances, cortisol helps manage your blood pressure and ensures you have enough glucose in your bloodstream to keep your energy up. When it surges under stress, it also quiets other systems that aren't immediately necessary for survival. Now, once the crisis is averted—the car drives by and you realize you're safe—the adrenals stop producing as much of these hormones, and your heart rate, breathing, and other systems go back to their normal resting state. You move on.

This effective stress response was designed across eons of evolution to flare up only in response to true moments of danger. Unfortunately, the brain—our master threat detector, which ultimately triggers the adrenals via the pituitary gland—reacts to modern dangers in the same way it would react to matters of life and death. That means short-term stressors, such as work deadlines or traffic, can trigger your threat detector. And long-term stressors, such as a bully at school or a sick sibling who makes our home feel unsafe and captures every ounce of our parents' resources, also provoke a fight-or-flight response. Over time, our bodies can become chronically inundated with these stress hormones.

Just as a normal immune reaction can go out of control and cause chronic autoimmune inflammation, an unregulated stress response will result in chronically elevated cortisol, or hypercortisolism. Too much cortisol can have serious side effects, like suppressing our immune system so we're more susceptible to illness or causing us to overeat and gain fat because our body thinks it needs to store energy to combat the threat. There's a large amount of literature associating hypercortisolism with depression. It can cause severe insomnia, because the adrenal glands, responding to the brain's anxiety or stress-related arousal, keep producing cortisol instead of winding down their production at night like they're supposed to. Even if you can fall asleep, elevated cortisol can affect the quality of your sleep so you never feel rested. Those times when you wake up in the middle of the night with your mind racing and you can't stop thinking about everything you have to do? You can probably thank cortisol. Long term, it can cause other major side effects like osteoporosis, cataracts, and diabetes.

Eventually, the adrenals tucker out and stop producing enough cortisol to keep your body working at optimal levels. I call this functional hypoadrenalism. Like a flashlight with a draining battery, the body is functioning despite lower than ideal levels of cortisol, but it could work much, much more effectively if the adrenal glands were in top shape. Many things can cause hypoadrenalism, the most common being autoimmunity. But chronic or persistent infections can do it, too, as can exposure to environmental toxins. Excessive amounts of exercise, such as training for a marathon or engaging in intense workouts after a long period of little movement, can tax the adrenals as well. And finally, chronic stress—both physical and emotional, with accompanying depression, anxiety, or mood dysregulation—is known to cause adrenal dysfunction.

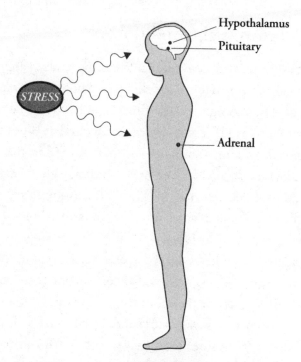

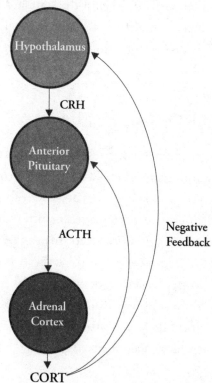

Hypothalamus
Pituitary
Adrenal
Axis

CRH
Corticotropin
Releasing
Hormone

ACTH
Adrenocorticotropic
Hormone

CORT
Cortisol

Another Controversial Condition

As we've previously discussed, conventional medicine tends to approach illness in a black-and-white paradigm—sick/not sick, normal/not normal, functioning/not functioning. Because functional hypoadrenalism—frequently referred to as HPA-axis dysregulation, and at times more controversially called "adrenal fatigue"—doesn't fit within that paradigm, many doctors dismiss its existence, declaring that a patient can be either adrenal deficient (a condition called Addison's disease) or adrenal sufficient. These doctors rely on standardized blood tests that assess cortisol levels in the blood to confirm their diagnosis. And yet I have patients who suffer symptoms of adrenal insufficiency—particularly fatigue and depression—whose cortisol blood-test results show that they are *just* above the arbitrary line dividing an adrenal sufficient patient from an officially adrenal deficient one.

Are their adrenal glands healthy? They're not. What can register as a "normal" cortisol level in the standard testing may not look that way when the same patient is put through a different test, one that too many doctors don't do even though it's a painless and relatively easy procedure.

Cortisol Testing

The typical adrenal blood test is performed twice during the day, once in the morning and once in the late afternoon. Part of the adrenals' job is to give you energy, so cortisol levels should start at their highest in the morning, then decrease until they get to their lowest point at night, allowing you to sleep. Instead of screening the blood for a snapshot of a patient's cortisol level at only one or two moments, I prefer to get a picture of my patients' cortisol levels throughout the day. By testing cortisol levels using salivary samples four times a day—morning, noon, around five p.m., and before bed—we can get a much more accurate sense of how the adrenals are performing.

Traditionally, salivary cortisol tests are used to check for Cushing's syndrome, or excessive levels of cortisol. But increasingly, researchers are finding that this easier, less invasive test correlates with serum cortisol levels, and is also useful for adrenal insufficiency. That's because salivary tests, unlike blood tests, measure free cortisol. You'll recall from the last chapter that sometimes the thyroid hormone in the bloodstream does not function optimally because its active part is too tightly bound to a carrier molecule. This can cause the body to respond as though it has a thyroid deficiency even though blood tests say the hormone levels are normal. The same thing can happen to adrenal hormones when they are bound up by a carrier molecule and thus made unavailable—the cortisol blood levels look normal, but the body reacts as though there's a deficiency because the cells aren't getting the information they need to function optimally. When tests reveal this is happening, I can tailor a regimen of herbs and nutrients to help quiet high cortisol or boost low cortisol.

Regulating the HPA Axis

By the time I brought my concerns to their attention, Charles's parents had watched their older autistic son, Will, make such big strides under my care that I'd earned a lot of trust. Once they became aware that their younger son was hurting, they made an appointment for him to see me. His salivary cortisol test showed his adrenal levels were low, a result, I suspected, of an adrenal system that had buckled under years of strain.

I believed Charles needed the support of a therapist so he could have a safe place to express his feelings and be assured that there was someone on his side who could help guide him through what must have felt like a minefield of a life. His parents needed to be a part of that therapy, too, so they could gain awareness of the unhealthy dynamics that had developed and learn to communicate with their child and respond to his needs. However, while therapy would surely make a huge difference in Charles's ability to cope and his overall mood, I feared it might not be enough to keep him mentally or physically healthy down the road.

Research shows that many children with pediatric depression also have a dysregulated HPA axis. The HPA axis—which like many of the organs and systems in a young, growing body shows high levels of plasticity, meaning it's easily malleable because it's still developing—helps shape our stress response throughout our lives. Studies have shown that children who suffer trauma in childhood, which could be due to anything from random adversity to parental abuse or neglect, are at higher than normal risk of developing PTSD, bipolar disorder, and depression in adulthood. Researchers suspect that, in addition to being immunosuppressive, sustained stress in the early years of a child's life can prime the body to emit a heightened stress response even when the child is no longer in the same stressful environment or whatever was triggering the stress is gone. It's like the nervous and endocrine systems get locked into an overreactive pattern that makes it hard for a person to respond appropriately to life events when they get older. In sum, a child with an untreated dysregulated HPA axis is at high risk of becoming an adult with a dysregulated HPA axis, even when the circumstances that led to the dysregulation have been neutralized or left behind. Fortunately, these regulatory systems can be repaired.

Treatment

Now that Charles was on his way to getting the emotional and psychological help he needed, we could start working on supporting his body's dysfunctional adrenal status. For this I rely on nutrients such as vitamin C, B-6, zinc, magnesium, and pantothenic acid, and herbs such as ashwagandha and rhodiola. His tests had not only showed sub-optimal adrenal levels, they'd also revealed Hashimoto's thyroiditis. This wasn't a surprise—the two often go together, and as we've discussed, Hashimoto's is strongly associated with stress.

But focusing solely on Charles's adrenals or thyroid wasn't going to heal him. We had to address the whole picture, and that included mental and emotional support. Despite his covert depression, Charles's therapist didn't feel he needed a prescription for antidepressants, and after a few weeks of family therapy—bolstered by good nutrition, hormones, and herbs—he started to come out of his shell.

I should caution that not all of my depressed young patients fare as well as Charles. There are times when parents bring me their children in the hopes I can help find a way to lower their dosage of an antidepressant or wean them off medication altogether. Sometimes I can, but sometimes I can't. Sometimes kids really need these medicines, which can make a tremendous difference in a child's moods and life. However, for kids who have seen some mood and cognitive improvement with medicine but have plateaued, adrenal support may get them over the hump. And sometimes, as with Charles, getting the adrenals to function properly helps all the other treatment protocols (including thyroid) gain traction and move the patient more quickly toward better physical and mental health.

Reactive Hypoglycemia

I put Charles on a diet of small, frequent meals comprised of lean proteins, vegetables, good fats, and low refined carbohydrates. Studies have shown a connection between a high glycemic index diet— one filled with refined carbs like white flour, sugar, and pasta—and anxiety and depression in teens. Proposed mechanisms that explain this connection, in addition to blood sugar imbalances, include inflammation, oxidative stress, and micronutrient deficiencies (folate, zinc, and magnesium). It's not surprising, then, that some studies have shown that switching to a Mediterranean-type diet replete with vegetables, fruits, fish, nuts, and olive oil may reduce the risk of developing depression.

Many of my patients who fall on the spectrum of adrenal dysfunction, however, do have evidence of low blood sugar. Reactive hypoglycemia—the technical term for this condition—occurs when the body's inappropriate or dysregulated insulin response to sugar and refined carbohydrates causes a significant drop in blood sugar.

Around 90 percent of diabetics who take insulin have experienced a bout of hypoglycemia. But nondiabetics can be hypoglycemic, too. It's considered rare, unless like me you work in a practice that specializes in treating patients experiencing chronic, taxing illnesses. Kids suffering from nondiabetic hypoglycemia struggle to cope with even mild stress and are often erratically emotional as their glucose levels spike and fall throughout the day, making it difficult for them to calmly handle normal life events. Since glucose is the brain's primary fuel, it's no surprise that tween and teen hypoglycemics often present with psychiatric symptoms such as an inability to concentrate, irritability, and volatile mood swings. Reactive hypoglycemia has been connected to psychiatric symptoms like insomnia, phobias, forgetfulness, emotional volatility, schizophrenia, and manic depression; neurologic symptoms like fainting, tremors, dizziness, and headaches; and generalized symptoms such as fatigue.

Like persistent Lyme disease, both functional hypoadrenalism and reactive hypoglycemia are conditions that many doctors claim don't exist. They do exist, and they often show up as a package deal in many of my patients, such as Ryan, whose bearded face could have appeared in the dictionary next to the word "strapping."

The Nature Lover Who Couldn't Go Outside

Ryan was never happier than when he was outdoors, preferably miles from civilization—but when I met him, he was so bent with fatigue, headaches, and severe anxiety that he was almost always afraid to leave the house. Ryan was twenty-two years old when he first came to my office, just out of his teen years, but is a clear example of how one inadequately treated condition can produce a domino effect.

In 2013, Ryan had been diagnosed with Lyme and treated with

the typical four-week course of the antibiotic doxycycline. He felt better afterward, but then his symptoms returned. Unfortunately, since he had already received the standard treatment, no doctor would give him another dose of antibiotics. The Lyme was gone, they told Ryan; he must have post-treatment Lyme disease syndrome. There was nothing more they could do. Ryan couldn't accept that explanation. He could hardly walk fifteen minutes, much less complete the five-mile hikes and half-marathons that were a regular part of his weekends. He had swollen glands, chills, and low-grade fever, and was sleeping up to sixteen hours per day on the weekends. He'd never felt so terrible in all his life.

Two or three months later, he was sitting in my office asking for help. Though he had an easygoing manner, he said he felt like he was always running at full speed, unable to relax. He swore he wasn't under any undue stress aside from concerns over his health, yet he was in a constant state of anxiety and had trouble concentrating. He mentioned irritability and shakiness in between meals, but assured me he'd been that way since he was a kid. It was normal for him, even if he felt like the tendency had gotten worse since the Lyme disease diagnosis. The irritability and the lifelong shakiness made me suspect he'd been living with reactive hypoglycemia for a long time, and that it had been exacerbated by the Lyme.

Sure enough, when his glucose test came back it revealed low blood sugar. I prescribed another round of antibiotics for the Lyme and put him on a hypoglycemia diet similar to the one I placed Charles on—frequent small meals, low if any sugar, lots of protein, good fats, and vegetables.

We thought we'd gotten a happy ending: all of Ryan's symptoms eased, he didn't get back in touch, and I considered his case closed.

But then, about seven months later, Ryan was back. "I feel like I'm dying," he told me. He'd attempted to hike the Appalachian Trail, but repeatedly he'd find himself deep in the woods, four or five miles away from any road, and suddenly overwhelmed with all the symptoms of a panic attack—chest pain, anxiety, dizziness, light-headedness, and adrenaline rushes. The one place in the world that had always provided Ryan with a sense of peace no longer felt safe.

Solving the Puzzle

Normal fasting blood-sugar levels should hover somewhere between 70 and 90 mg/dl (milligrams of glucose per deciliter of blood), with higher levels in the low to mid-100s if not fasting. When I tested Ryan's blood sugar, it registered 44 mg/dl. In addition, lab tests showed evidence of low hemoglobin A1C (a measure of blood sugar over the past two to three months), consistent with hypoglycemia, as well as evidence of hypocortisolism, consistent with adrenal dysfunction.

So why did hiking, Ryan's favorite activity, seem to make everything worse? Part of the issue was the physical exertion, but it was also what he was eating to keep up his energy on the trail. Hikers can't carry pounds of lean meat, fruits, and vegetables in their backpacks. They're dependent on compact, lightweight, calorie-rich foods like protein bars, trail mix, oatmeal, and granola. Foods that were high in carbs and sugar would make his blood sugar spike and then drop dramatically. Further, his low adrenal function would impair his body's ability to regulate those highs and lows.

Treatment

For the short term, I put Ryan on the antidepressant Celexa and a benzodiazepine, Klonopin, to help control the panic attacks until we could get him back to good physical health. I gave him three homeopathic formulations bundled together in a "stress buster kit" to help him cope and rebound from his stress.

It was a slow process, but over time, Ryan's Lyme disease–induced fatigue and headache gradually improved on antibiotics, including intramuscular Bicillin. He noted much improvement in all his other symptoms when he strictly adhered to the hypoglycemic diet and various supplements. These included L-glutamine, methylated B vitamins, magnesium taurate, vitamin C, GABA/L-theanine, chromium, biotin, zinc, and my own combination formulations that contain several of these, as well as herbs such as *Magnolia*, *Phellodendron*, and ashwagandha. These were used to help support and balance his adrenal function and carbohydrate metabolism. Ryan also sought out the help of a psychotherapist, who gave him the tools to avoid his panic attacks.

The better Ryan stuck to his diet, the less often he hit those lows that made him feel shaky, anxious, and "off." And on the occasions when he did start to feel that way, he now knew how to talk himself down and stay calm until he could eat something and get his blood sugar back where it needed to be.

In retrospect, Ryan believes that his hypoglycemia probably played a big part in the social anxiety he suffered in school. He remembers sweating a lot, experiencing constant upset stomachs, and often feeling "groggy, crummy, and anxious," which had led to low self-confidence and difficulty forming good relationships. Today, Ryan says he's strict with his diet and feels the best he's ever felt about himself. He's off all medications and back on the trails, not just in the Adirondacks but in all the national parks, anywhere he can find solitude, peace, and mountainous terrain.

The Mono That Wasn't Mono

Michelle was another patient of mine who experienced tremendous relief once we got her hypoglycemia and adrenal system under control. If you ask her mother, Della, to talk about what it was like to watch illness rob her daughter of her friends, favorite activities, and her education, she falls silent. It's hard to get the words out when you're trying not to cry.

A cheerleader, softball player, and competitive dancer who practiced five nights a week while still regularly landing on the high honor roll, Michelle was a typical if somewhat type A fourteen-year-old with wavy brown hair and almond-shaped eyes. In January 2015, during the last few days of winter break, she attended a sleepover with a few other girls from her eighth-grade class. Shortly afterward, she and another girl from the sleepover both fell ill. The other girl was diagnosed with mononucleosis, and Michelle's symptoms—extreme fatigue, migraine headaches, muscle weakness, and sore throat—were so similar that Della and her husband, Ben, assumed that's what their daughter had, too. But while the other girl got better, Michelle did not. Yet she had no fever, so when school started, they insisted she return.

But things did not go well. Along with the aforementioned mono-like symptoms, Michelle had joint pain, nausea, and dizziness, along with heightened irritability, anxiety, and what appeared to be some brain fog or memory loss. Sometimes she got so light-headed she fainted. Their local pediatrician ran the standard mononucleosis test, which came back negative, as well as a Lyme test, which was also negative. None of Michelle's bloodwork nor a brain MRI showed anything that could be causing her symptoms. Stumped, the pediatrician recommended that Della and Ben bring Michelle to Children's Hospital of Philadelphia (CHOP), considered one of the best children's hospitals in the country. There she underwent another battery of tests, including a tilt-table test, during which the doctor took Michelle's blood pressure every few minutes while she lay on a table that tilted her at varying degrees, as well as when she was lying down and standing up. At one point, Michelle was falling over, unable to hold herself up.

After that test, the doctors at CHOP knew why. Michelle had postural orthostatic tachycardia syndrome (POTS), an autonomic nervous system dysfunction in which the body can't regulate heartbeat and blood pressure when a person goes from lying or sitting to standing. Symptoms generally include an excessively fast heart rate, low blood pressure, dizziness, and fainting. She also had joint hypermobility syndrome. Joint hypermobility is a fancy term for someone who might describe themselves as double jointed. In most people this inherited condition is benign, and in Michelle's case, maybe even beneficial—it was probably one of the gifts that made her such a good dancer. But in some people, joint hypermobility causes joint and muscle pain, easily dislocated joints, and dizziness and faintness, among other symptoms.

The recommended treatment was for Michelle to increase her fluids and salt intake. She was supposed to drink 60 fluid ounces of water per day, but she found it hard to drink that much. She was also prescribed a steroid called fludrocortisone, a synthetic mineralocorticoid, which helps the body hold on to salt, but that just made her sicker. Meanwhile, Della had Michelle tested for Lyme again through a different lab in California. This time, the results came back positive not only for Lyme, but for several other tick-borne coinfections. No one had ever noticed a tick or a bull's-eye rash on Michelle, but the presence of an IgM antibody suggested that the infection was relatively recent, perhaps from that autumn, although at times we do see persistently elevated IgM titers. It was certainly possible that the untreated Lyme had triggered the POTS, but no one could say for sure.

It was now May, five months after Michelle had first fallen ill. Upon hearing about the Lyme diagnosis, Michelle's pediatrician had immediately put her on the antibiotic doxycycline, but Della soon sought out a Lyme specialist, who kept up the doxycycline but added Mepron and several other antibiotics to deal with the coinfections. Michelle was still being battered by a host of physical symptoms typical of POTS and Lyme—near-fainting and recurrent fainting; chronic fatigue, yet chronic insomnia, too; brain fog; cognitive dysfunction; headaches; joint aches of the knees, elbows, and thumb; air hunger; occasional chest pain; nausea with occasional vomiting; gastroesophageal reflux (GERD); and sound sensitivity. Della had noticed a drastic change in her daughter's personality, too. She'd always had OCD-like tendencies, but over the months it had worsened. The normally gregarious, confident girl became intensely moody, defiant, and anxious. Her school had refused to let her attend for half days, so she was being tutored at home an hour or so per day, which was about as much as she could handle. She'd stopped dancing. At times she vomited several times per day, and the constant feeling of nausea was debilitating. Michelle's whole life had shrunk to the confines of her house. Della started hunting for someone who might be able to coordinate a treatment for all of Michelle's symptoms. It took awhile, but finally, in March 2016, she came to see me.

If the now fifteen-year-old teenager had been sitting in a chair between her parents when I walked into my office, instead of lying down on my couch, I would have called this a picture-perfect family. Though underweight and sporting dark circles under her eyes, Michelle was a beautiful girl; Della and Ben were both strikingly good looking and extremely kind, gentle people. I talked for a long time with the adults, and then turned my attention to their daughter.

Sometimes teenagers will clam up when I focus on them. Not Michelle. She wanted to tell me everything and struggled to maintain her composure as she described how her friends had rejected her, including her best childhood friend, because she kept flaking out on them and wasn't much fun to hang out with. She told me her dance teacher didn't believe that she was as sick as she claimed, and that the school had accused her of faking her illness. She couldn't do any of the things she liked to do anymore. Her mother, who worked for the school district, had been put in the awful position of trying to hold on to her job while fighting for her daughter's right to continue her education. As she spoke, she'd cry, pull herself together, then cry some more. I felt for her, for all of them. Part of the strain on all of these families is knowing how many people don't believe what they're going through. When they finally get a chance to talk about what's happened to them, it can unleash a wave of emotion that's hard to control. The best thing I can do is listen and gain their confidence.

Once Michelle was ready, I conducted a physical exam, and as frequently occurs in the cases that make their way to my office, the exam and lab results revealed a host of issues that often seem to cluster in patients carrying Lyme disease. Extending on both sides of her body from her hips to her buttocks were purple striae that could have been mistaken for stretch marks if she were heavier, a physical sign suggestive of Bartonella. There was the POTS, of course, which is often associated with tick-borne illness. And then Michelle's thyroid was slightly plump. Subsequent testing would reveal Hashimoto's, also commonly associated with tick-borne illness. All of this in addition to the fact that despite being treated for Lyme and POTS she wasn't getting any better made me wonder—was it possible that the mono-like illness that had struck almost a year earlier was a viral infection that had triggered autoimmunity, paving the way for these conditions? Her blood tests showed positive IgM and IgG antibodies for Coxsackievirus, suggesting it was still a somewhat persistent active viral disease.

There's overlap between the symptoms of POTS and those of adrenal dysfunction. In addition to adrenaline and cortisol, the adrenal glands release aldosterone, a hormone that lets the kidneys regulate the body's water and salt balance. A lack of aldosterone blocks the kidneys' ability to do their job, including retaining sodium, which leads to low blood pressure, causing people to feel dizzy and faint. Sure enough, when I conducted the salivary adrenal test about a week after her first visit, Michelle turned out to be cortisol deficient in the morning, yet high at night, exactly the opposite of what you'd want to see.

And then I ran one more test. The crying, defiance, irritability, and increased anxiety could all be explained by the mere fact that this girl felt sick day in and day out, with no relief in sight. She'd lost everything that mattered to her except the love and support of her family. Anyone would cry. But something else Della had said made me wonder if there wasn't more to the abrupt personality changes she'd seen in her daughter. She'd said that Michelle's bad moods and emotional volatility tended to spike when she didn't eat much during the day. Again, understandable. Most of us get cranky when we don't eat. But she also said that Michelle craved sugar. In fact, Michelle's whole diet was dominated by carbs. At the time she started seeing me, her diet was heavy in pasta and pizza, as well as bread, sweet cereals, and sweetened yogurt. She drank milk and apple juice, and ate raspberry ice every day. Further, she was eating more than she had been, thanks to a proton-pump inhibitor called lansoprazole that controlled her reflux, but she was eating almost nothing but sugar and refined carbs. The constant funneling of sugar into her system meant that her body was in a perpetual state of highs and lows as it tried to regulate her glucose levels.

To recap, Michelle had been on two months of antibiotics to treat Lyme disease. Her previous doctors had prescribed more fluids, more salt, and fludrocortisone for the POTS, but she hadn't been able to fully keep up with the first two, and hadn't been able to keep down the mineralocorticoid. On top of that, she was feeding herself almost nothing but refined carbs and sugars. And on top of *that*, she was constantly nauseated and under massive mental and emotional strain from losing her friends and her spots on her softball and dance teams, and from wondering if she'd ever regain her equilibrium, both literally and figuratively.

Treatment

There were many possibilities as to what could be contributing to Michelle's POTS, but before we did anything we needed to rebalance her system. The day of Michelle's first visit, we attacked the POTS with an intravenous solution of sodium chloride, adding a dose of vitamin C with magnesium, the latter good for modulating the immune system and revving up one's energy. In addition, I added a couple of homeopathic remedies specifically to help with her GI complaints, especially nausea, as this can make complying with medications and supplements extremely difficult. Thankfully, she was what I call a homeopathic responder, and the nausea eased.

Next, we had to get her diet under control. Several studies suggest that high glycemic eating habits can be associated with depression and other psychological symptoms, and in a published case report, a fifteen-year-old suffering from generalized anxiety disorder saw marked diminishments in anxiety and hypoglycemia symptoms when her diet was shifted from one primarily consisting of refined carbohydrates to one made up of protein, fiber, and fat. I was sure that if we could balance out Michelle's roller-coaster blood-sugar spikes and subsequently help shore up her emotional resilience, she'd see faster improvements.

Reactive hypoglycemia is relatively easy to treat with a strict healthy diet of frequent small meals that replaces empty carbs and sugars with lean protein, vegetables, nuts, and good fats. Paradoxically, reactive hypoglycemia can also be tough to treat, not because adjusting one's diet is so complicated, but because teens often just don't want to cooperate.

Would Michelle comply?

Buy-In Matters

If there's one thing that will get a reluctant teenager to come around to a treatment they've resisted, it's physical discomfort. If they've got symptoms like frequent gas, mushy, stinky stools, acne, or halitosis, they'll likely cooperate because they'll do anything to make those go away. But with psychiatric symptoms, a child can get so used to their low energy that they don't realize how much better they could feel. Michelle had undeniable physical symptoms that were putting a wedge between her and her friends and teachers, but she couldn't recognize how her emotional volatility and irritability, caused by low blood sugar, had affected her relationships.

Regardless, this is one of those instances where a child's buy-in matters. In many of the conditions we've discussed, medicines and nutritional supplements have proven extremely helpful. Swallow a thyroid supplement and your thyroid function starts to improve; take an antibiotic with a chaser of herbs and you can get rid of Lyme disease symptoms. However, treating low blood sugar (and allergies, which we'll discuss later) is more of a long game, requiring a tween or teen to think about diet and lifestyle and make a conscious decision to change.

This is when I have the serious conversation with my patients, explaining that they've now come to a point in their lives when they need to take responsibility for their own health. I can't be with them all the time to keep them in line. Their parents can't hover over them to make sure they make good choices. I treat them like the young adults they're becoming, and they seem to appreciate it. Rare is the teen who will make changes just because an adult told her to, but the majority of the teens I treat do step up and start taking charge of their health. Michelle was one of them. Fortunately, she stuck to her diet, which was essentially a Paleo regimen: protein, vegetables, and good fats, with occasional whole grains. No gluten, no dairy for at least a few months. No more sugar cereal, that's for sure.

To combat the Coxsackievirus, I prescribed a couple of natural antimicrobials, including Lauricidin, derived from coconut oil, and olive leaf extract, as well as the nutrients vitamin D and vitamin C, which would also help provide immune system support, and finally, transfer factors, which help the body defend itself by conferring some immunity against certain infectious agents.

Now it was time to get Michelle's adrenals back in shape. We started with a low dose of hydrocortisone to be taken in the mornings when she seemed to be low on cortisol, and added supplements like glutamine, zinc, and magnesium, as well as my own combination formulation consisting of high doses of chromium and biotin, to help reduce her sugar cravings and improve her carbohydrate metabolism (see appendix for details). We also discovered she had impaired methylation capability (a condition we'll discuss in chapter 9), so I put her on methyl B-12 and methylfolate to correct for it. Was it contributing to the POTS, adrenal insufficiency, or reactive hypoglycemia? I can't say. But when you're dealing with chronic illnesses that don't have one known cause, the most efficacious thing you can do is balance the system as best you can. You find the deficiencies and insufficiencies, you find the imbalances, and you correct them. Methyl B-12 is important for metabolism and detoxification, and can often help combat fatigue.

Several months later I noticed that the purple Bartonella striae on Michelle's hips were deepening. She tested negative for Bartonella, but then I was left with two choices: cross my fingers and hope things didn't progress, or knock out a potential infection before it could do more than a little cosmetic damage to Michelle's skin. I chose to knock the bug out. After three months of antibiotics she was feeling significantly better, and the dark purple Bartonella tracks faded.

Relief

It's incredible how much difference a treatment regimen comprised mostly of dietary changes and a few herbs and nutritional supplements can make in a person's life. I suppose that's one of the main reasons my patients tend to follow my recommendations—when they do, they often feel better, and then they want to see just how much better they can feel.

Michelle had been sick for over fourteen months by the time I started treating her, so it took a while for her to fully recover, but Della reported that Michelle's fatigue lifted almost immediately, as did the nausea, which made a huge difference in her morale and ability to comply with the rest of her treatments. Her joint pain and muscle weakness ebbed. Over the course of her junior and senior years of high school, her health stabilized and she was able to participate more and more in ordinary teenage activities.

Michelle is a freshman in college now, attending school about an hour north of her family home. She's a healthy eater, sticking to nondairy beverages, very little sugar, and organic options. She avoids red meat, not because I advised her to but because she says she notices a difference when she does. Same with gluten, though she's not strictly gluten-free. She continues to take a probiotic and several supplements to keep her immune system strong. There is no cure for POTS, but her symptoms are manageable now. And the girl who could once hardly sit up, much less stand, earned herself a spot on her college cheerleading squad.

Michelle, Charles, and Ryan got better because we did the detective work to find the physiological causes of their symptoms and treated them accordingly—instead of just treating the symptoms themselves. It's unfortunate that many doctors still debate whether these conditions even exist. My hope is that this will change as word gets around that it's possible to successfully treat and heal complicated patients like these.

COULD YOUR CHILD HAVE ADRENAL DYSFUNCTION?

Does your child have symptoms that are worse when he doesn't eat for a while, such as irritability, shakiness, fatigue, or light-headedness?

Do these symptoms improve after meals?

Does your child crave sugar and sweets?

Is your child's mood or anxiety affected by what they eat, including sugar, refined carbohydrates, and sweets?

Is your child chronically low energy? Do they complain of weakness?

Does your child suffer from brain fog?

Does your child complain of dizziness or light-headedness upon standing?

Does your child have trouble sleeping?

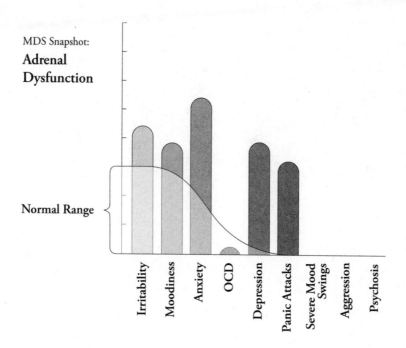

MDS Snapshot:
Adrenal Dysfunction

Normal Range

Irritability · Moodiness · Anxiety · OCD · Depression · Panic Attacks · Severe Mood Swings · Aggression · Psychosis

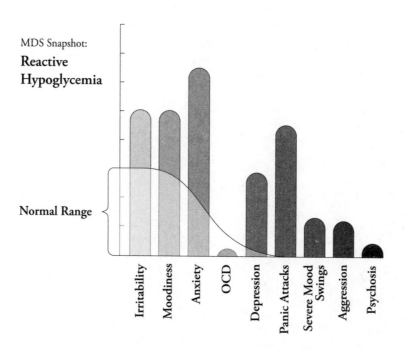

MDS Snapshot:
Reactive Hypoglycemia

Normal Range

Irritability · Moodiness · Anxiety · OCD · Depression · Panic Attacks · Severe Mood Swings · Aggression · Psychosis

CHAPTER 8

Overreacting to Life
Allergies and Sensitivities

Everyone thinks they know what an allergic reaction looks like: red, itchy eyes and a runny nose that erupts within twenty minutes of entering a house with a cat; a scratchy mouth and throat after eating raw carrots; hives after taking a medicine. Triggered by anything from birch pollen to pet dander, peanuts to penicillin, more than 50 million people in the U.S. suffer from seasonal, indoor/outdoor, food, skin, and drug allergies every year. As you'll recall, an allergic reaction is an immune response in which the immune system mistakes a normally harmless antigen, such as a dust particle, as a threat and produces large numbers of IgE antibodies to combat it. When the immune system comes into contact with that antigen again, the IgE antibodies are ready and waiting to mount an attack, releasing histamines and other inflammatory molecules that can result in itchy or swollen eyes, drippy nose, asthma, rashes, and a host of other physical symptoms.

Food allergies are particularly common—it's estimated that 5.6 million children in America have them. That's about 8 percent, or one in thirteen kids. Eight foods are responsible for about 90 percent of food allergy reactions: milk, eggs, wheat, soy, tree nuts, peanuts, fish, and shellfish—with milk, eggs, and peanuts being the most common. However, just about any food can cause an allergic reaction.

And as it turns out, everyone doesn't actually know what an allergic reaction looks like. In fact, some allergies don't manifest with physical symptoms at all.

The Brilliant but Struggling Student

I appreciate all my patients, but Eden, a vibrant little girl with a tangle of brown curly hair, is a delight. Precocious and charming at age eleven, she's the kind of child who isn't intimidated by grown-ups, and is as interested in getting to know you as you might be in getting to know her. She was already that way when I met her three years ago, at age eight. Her parents brought her to me because although she had tested off the charts for intelligence, she was struggling in school. No one had noticed anything unusual in her behavior when she was in kindergarten and first grade, but now that she was in second grade and expected to sit still and complete schoolwork as opposed to engaging in free play, her teachers were becoming concerned. She couldn't finish simple worksheets. She seemed lost unless someone was engaging directly with her and reminding her of what she was supposed to be doing or where she was supposed to be. Her teachers also noticed that she couldn't sit still in her seat.

Lily and Ben, Eden's parents, noticed the fidgetiness at home, too, particularly at mealtimes, but they supposed it was normal for such a young child. And it wasn't as though their daughter's quirks were disruptive. So what if she was a spacey, dreamy girl who repeatedly got caught drawing pictures or talking to her stuffed animals or building towers out of shoes instead of getting dressed in the morning? Was it frustrating? Sure, but there are worse things.

As the school year progressed, however, Eden's grades kept sinking and her inattention was starting to get her into trouble as she constantly had to be redirected and refocused. Lily and Ben weren't surprised when a neuropsychologist diagnosed Eden with ADHD, but they were uncomfortable following the doctor's suggestion to medicate.

I'd met Eden's parents a few years earlier when they'd brought in Eden's little sister, Dahlia, after she was diagnosed with an autism spectrum disorder. Under my care, Dahlia had progressed extremely well and in fact was about to leave her special needs classes and join a general education class in a new school for the first time. Lily and Ben hoped I might be able to offer some kind of natural regimen that could help improve Eden's school performance without putting her on stimulants.

I knew from interviewing both Eden and her parents that the girl had your garden-variety seasonal allergies that flared up in the spring and fall, as well as allergic asthma, for which she used an albuterol inhaler, a nasal spray, and an antihistamine as needed. Still, when testing for allergies, I generally use two different techniques. One is a radioallergosorbent test, or RAST test, which requires a blood draw and measures the amount of IgE antibodies that appear when an allergen is added. The other is an intradermal skin test. Unlike the more common scratch test, in which the skin on the outer arm or back is lightly pricked with fine needles (called lancets) and a minuscule amount of allergen is applied to the area, the intradermal skin test involves injecting a very small amount of allergen into the dermal area of the skin, creating a wheal (a small raised area). If the wheal swells and turns red a certain amount, that's a positive test for an allergy. The greater the reaction, the greater the allergy. Intradermal skin tests have the advantage of being relatively painless (requiring only a small prick) and delivering fast results right there in the office. They're also more accurate than skin prick tests. I've seen a large number of cases in which a skin prick test is negative for food allergies, but the blood test or intradermal skin test comes back positive. Blood tests are a great backup because you can test a lot of foods with just one stick.

I ran an intradermal skin test on Eden for milk, wheat, and corn, the three allergens that, in my experience, often seem to go hand in hand with children's behavioral issues. Eden immediately tested positive for a corn allergy. Honestly, that's a tough one, harder to eliminate from a diet than even wheat or gluten. Corn is in *everything* (you can read all about why and how that happened in Michael Pollan's seminal book *The Omnivore's Dilemma*). It would take major dedication for Lily and Ben to make sure their daughter's diet was corn-free. But Lily and Ben themselves were very bright, natural overachievers. If anyone was up to the challenge, it was them.

It took painfully detailed research to remove corn from Eden's diet because the FDA-enforced Food Allergen Labeling and Consumer Protection Act (FALCP) only "requires that food labels must clearly identify the food source names of any ingredients that are one of the [eight] major food allergens or contain any protein derived from a major food allergen." Corn isn't one of the eight major food allergens, so it doesn't have to be listed on food labels and can hide in the most surprising places. You might feel it's a safe bet to simply assume all processed foods contain corn, but would you think to check the ingredients for, say, vanilla? Yet "pure" vanilla extract can contain corn syrup (which is usually listed) and is made with 35 percent ethyl alcohol, which can be derived from corn (a detail that might not be listed). Corn can be found in toothpaste, makeup, and the glue on the back of envelopes. In Lily's words, "it's insidious."

Even some of the supplements I use, which are generally hypoallergenic and free of gluten, dairy, soy, and eggs, wouldn't work for Eden because the capsules contain trace amounts of corn. Every single school lunch had to be homemade. All the family's bread was now home-baked. Halloween was a particular challenge, but Lily and Ben handled it by making special corn-free candies and treats and handing them out to Eden and all her friends. Lily felt like she was spending all her time in the kitchen, and she had a full-time job and two other small children to raise!

But the effort paid off. The next time I spoke with Lily, about four to six weeks later, she reported a rapid, "major shift" in Eden, with marked improvements in her attention, fidgetiness, distractibility, and even her mood. The girl seemed to have more energy, too, complaining less of fatigue and neck pain. All of Eden's teachers reported a positive change, and her grades started to climb. Since then, she's done extremely well in school, eventually even dropping the tutoring she used to require in math. The only extra help she gets anymore is from an executive function coach who is working on teaching her good organization and planning habits.

It's heartbreaking to think of all the children who start school with enthusiasm and joy, only to discover it's a place that seems to draw out the worst in them and invite criticism. Think of the students who get branded as "difficult" or "challenging," their inattentiveness mistaken for a lack of intelligence. These labels often stick and can limit adult expectations, as well as the kids' self-esteem and aspirations. After enough time, a child who's constantly corrected and disciplined, who gets used to being told that their behavior is unacceptable, will start to think of themselves as a bad kid, a screwup, a loser.

What if we were to find out that these kids' problematic behavior or inattention was simply a reaction to a glass of milk, a piece of bread, or a banana? What if we could make a difference in their academic experiences, and even their lives, just by eliminating a few key foods from their diet?

Back to the Gut-Brain Connection

We know that sometimes symptoms that appear neuropsychiatric, like depression, are in fact a result of inflammation. Since an allergic reaction is actually an inflammatory response, we shouldn't be surprised that multiple studies have found a connection between allergies and psychiatric disorders. One population-based study found that people with allergies had a higher risk of developing depression, anxiety, or bipolar disorder than those without. Cow's milk has been shown to trigger increased behavioral and emotional disturbances in mice sensitized to dairy, and doctors have seen positive results from elimination diets in patients suffering from treatment-resistant depression, indicating that the covert reaction to certain foods may have been contributing to a resistance to their medications. Consistently, elevated levels of IL-6 are found in people suffering from major depression and anxiety. IL-6 is one of the inflammatory cytokines responsible for loosening the tight junctions of the intestinal wall, allowing inflammatory molecules into the circulation, where they can then travel to the brain (see the gut-brain-immune axis in chapter 2). It can be produced in response to some medications, stress, infections—and allergies.

According to the CDC, the number of children with food allergies rose 50 percent between 1997 and 2011, and the rates continue to rise. No one is really sure why, and the research on how to prevent allergies continues to evolve. For a time the best strategy seemed to be for parents to avoid introducing allergens into their infants' diets, but recent studies suggest the opposite is true, and that offering allergenic foods to healthy infants can have protective effects.

Others postulate that the rise in childhood allergies may be due in part to our overreliance on antibacterial hygiene products that don't allow the body sufficient exposure to microbes or parasites; antibiotic use; C-sections and the resulting altered microbiome; obesity; vitamin D deficiency; and environmental factors such as chemicals, pollution, and heavy metals that compromise the immune system.

But the theory that's currently garnering the most support is that the rise in allergies could correspond with the rise in the incidence of intestinal hyperpermeability, or leaky gut. Many of the above physical and environmental factors, as well as stress, can contribute to this condition. It remains unclear whether it's the leaky gut allowing large proteins to enter the circulation that triggers the allergic immune response, or the allergic reactions themselves that loosen the tight junctions of the gut lining. Regardless, researchers and my own observations confirm that loss of the integrity of the intestinal lining plays an important role in susceptibility to food allergies.

Food Sensitivities

Often when a child's food allergy tests come back negative, their parents will continue to observe a pattern of what seem like allergic symptoms or discomfort each time their son or daughter is exposed to a certain food. In these cases, while the child may not be allergic to that food, they could have a sensitivity to it. Whereas an allergic reaction is specifically a near-immediate immune response to an offending antigen, a food sensitivity is different and can often manifest long after a person has been exposed to the offending substance, sometimes by several hours or even days. It's important to note that food sensitivity shouldn't be confused with food intolerance, which is a digestive disorder in which the body doesn't produce enough of a particular enzyme (such as lactose) to break down certain foods (such as milk) in the intestines, resulting in symptoms such as intense bloating, diarrhea, gas, nausea, vomiting, and heartburn.

"Food sensitivity" is not yet a formal medical diagnosis and can be met with skepticism, but ask anyone who suffers a reaction every time they eat a certain food despite testing negative for allergies—or the parent of such a patient—and they'll tell you it's a very real thing. And in my experience, even when sensitivities don't cause a full-blown allergic reaction, specific foods can cause uncomfortable physical, behavioral, and neurological issues.

Dairy, eggs, and wheat commonly elicit sensitivities. Gluten is another trigger for many people. While only 1 percent of the population is believed to have celiac disease, an autoimmune disorder that causes gluten to destroy the lining of the small intestine, initiating damage that can lead to pain, diarrhea or constipation, anemia, malabsorption, and malnutrition, millions of people without celiac have found relief from all sorts of ailments—including digestive issues, chronic fatigue, and swollen joints—by avoiding gluten. And in my practice, I've spoken with many parents whose children have struggled with mood swings, who swear that the effect of eliminating gluten from their child's diet has been near-miraculous. Tonya and Richard O'Meara are two of them.

The Lost Boys

Tonya loved being pregnant. Her joy was especially sweet knowing that her sister was also pregnant, with a due date just a few weeks ahead of hers. In the end, the age gap was even less than that. Her nephew, Emmett, was welcomed into the world only six days before her own son, Caleb. As Tonya and her husband, Richard, cocooned at home with their new baby, she fantasized about the shared birthday parties, vacations, and holidays the two families would enjoy together, and hoped the boys would grow up close friends.

As they navigated the ropes of motherhood at the same time, Tonya and her sister spent a lot of time sharing their experiences and observations with each other. The babies seemed to be following similar cycles and growth patterns, and for the most part Tonya didn't see any major differences between her nephew and her son until about six months.

Like many parents, Tonya had fun introducing Caleb to a world of new foods, starting with rice and single-grain cereals, then slowly progressing to fruits and vegetables, meat, and dairy. What wasn't fun were the diaper changes. Tonya knew her child's poops would change in consistency once he started solid foods, but the explosive diarrhea running up his back and out the legs of his onesies were more dramatic and frequent than she'd expected. To her surprise, this seemed to be a talent particular to Caleb; while Emmett certainly provided Tonya's sister and brother-in-law with some unpleasant diaper changes—his contribution to a bumpy airplane ride to Phoenix would go down as family legend—by the time he was a year old he was producing your standard diaper contents. Caleb's deposits, on the other hand, were so runny they required Tonya to change his clothes multiple times per day.

That wasn't the only difference Tonya and Richard noticed. Emmett sometimes got the sniffles, but Caleb was stuck in a never-ending rotation of ear infections or croup. Emmett started saying his first words; Caleb remained silent. Emmett was engaging with people and his family's pets, whereas Caleb seemed a little removed socially.

Though all children develop at their own pace, Tonya had a feeling something was up—and she was right. At eighteen months, Caleb was diagnosed with PDD-NOS—pervasive developmental disorder, not otherwise specified, a term that's no longer used, but that at the time was considered a subtype of autism, applicable to people who showed some symptoms of autism but didn't meet the strict criteria for a definitive diagnosis of frank autism (today Caleb would be diagnosed with autism spectrum disorder and placed in the "neurodevelopmental disorder" category of the spectrum).

Following the diagnosis, a child psychologist handed Tonya and Richard a booklet designed to help prepare parents for what to expect from a child with autism—which, in a nutshell, wasn't much. They were devastated, as the book warned them they shouldn't expect or hope for Caleb ever to lead a typical, socially engaged life. It assured them there was nothing they could do. Tonya and Richard's response was quick: Like hell there wasn't.

They plunged headfirst into the growing community of families determinedly seeking information about this little understood disorder. Immersing themselves so completely in the research, attending conferences, and seeking out experts is doubly impressive when you learn that at the same time they were also caring for another newborn. Only three months prior to Caleb's diagnosis, Tonya gave birth to their second child, a little boy they named Luke.

Fortunately, with the exception of suffering from frequent ear infections, Luke was an easy, happy baby, which made it easier for Tonya and Richard to focus on finding Caleb the help he would need to flourish. They had high hopes that speech therapy would help unlock his personality and ability to communicate, but despite weekly sessions, they didn't see much progress. Meanwhile, his diarrhea, croup, and ear infections continued. Tonya finally requested that the pediatrician test Caleb for allergies, but he refused, believing the child was too young to have allergies (a common misconception two decades ago). That's when she and Richard sought me out. I did conduct the tests and discovered Caleb was allergic to dairy, eggs, and peanuts. In addition, a pediatric gastroenterologist found that Caleb tested positive for antibodies to gliadin, a protein in gluten, though none of the other markers that would indicate celiac disease. Tonya and Richard immediately started removing the allergens from Caleb's diet, and for good measure, removed the gluten, too. Then they watched to see what, if anything, would happen.

Caleb's diarrhea resolved almost immediately. And just as immediately, Tonya and Richard saw other improvements. The ear infections stopped recurring. So did the croup. But best of all, whereas months of therapy alone had brought about little improvement, Caleb started talking. At first it was just a word here and there, but in time, Caleb was speaking in short sentences. He started to engage. He made so much progress that when he turned three, he was able to start attending preschool for a few hours per day, accompanied by a one-on-one classroom aide.

And thank God for that, because by then Luke was no longer a happy baby, and Tonya was losing her mind. Day in, day out, Luke screamed. He screamed if she put him down. He screamed when she dressed him. He screamed if he touched grass. He screamed at any loud noise. His meltdowns were epic, the kind that drew wide-eyed stares from everyone in view. Every parent has had to bear the withering glares of judgment when their child decides to have a temper tantrum in public; Tonya had to bear them every day, sometimes multiple times a day, because there was no way to predict what would set Luke off.

By the time Caleb was in first grade, he no longer needed an aide to accompany him during class and could communicate clearly, and there were few hints of the physical symptoms he'd suffered during his early years. Luke, however, was a miserable, cranky, angry, hypersensitive, and aggressive seventeen-month-old. Hoping that I could figure out what was causing his terrible moods, Tonya and Richard brought their younger son to me.

First, I identified a vitamin A deficiency and gut dysbiosis, which was no surprise given the number of antibiotics the child had been on for ear infections since early on. I also found that he had an immune deficiency, which likely explained his susceptibility to the ear infections. He hadn't had one in a whole month, however, nor a bout of croup—not since a visit to a different doctor who had suggested Tonya and Richard try taking Luke off dairy. That raised a flag. Luke didn't test positive for any allergies like his brother had, but if removing dairy could quickly resolve some of his physical ailments, maybe food was causing some of his other issues, too. Caleb was autistic and had had explosive diarrhea. Luke wasn't autistic, but he had an explosive personality. What if his moods were being affected by certain foods, similar to the way they caused GI symptoms in his brother?

A RAST test was negative for allergies, but an IgG test measuring the amount of food-related IgG antibodies in the blood revealed a sensitivity to casein, dairy, and peanuts, though not to gliadin. Allergists have long dismissed IgG testing because almost every food triggers an IgG response, so all it really means is that a person has been exposed to certain foods. Some doctors' concern is that people, and especially children, will limit their diets unnecessarily or develop an unhealthy relationship to food. To me, however, high levels of multiple food-related IgGs indicates that something is going on in the gut, likely intestinal hyperpermeability. The test doesn't offer me a diagnosis, but it does provide me with some good clues to follow.

The IgG levels in Luke's blood and his positive clinical response to going off dairy—diminishment in his recurrent croup and ear infections—made me suspect food could be at the root of his behavior problems. First, I treated him with antifungals and natural antimicrobials to address the gut dysbiosis, and then a series of supplements to support his immune system. We took the casein and peanuts out of his diet, too, and observed him closely. His brother was gluten-free so the amount of gluten in his diet was lower than average, but though we discussed removing it entirely, Tonya and Richard hadn't seen any reason to be strict about it since he didn't test positive for gliadin antibodies. Yet as Caleb's autism symptoms retreated further and further, they started to consider that maybe they should see what would happen if they put Luke on the same diet as his brother. It was 2006. Luke was now four and a half years old.

Just like that, with the gluten removed from his diet, he was a different kid. "An angel," his parents called him. He stopped crying and clinging to Tonya. She stopped getting phone calls from frustrated teachers. I saw the difference right away the next time he visited my office, where he sat through his exam calmly with no sign of aggression or a burgeoning temper. Though as Luke grew older he remained a little more emotionally volatile than the average child and retained some sensitivities, particularly to loud or grating sounds, however, these sensitivities didn't impede his school life or his ability to build relationships with friends and teachers.

Non-Celiac Gluten Sensitivity

Why does a gluten-free diet in particular so often seem to make a difference in the behavior of a child? Simple: it appears that gluten can have a direct effect on the brain, triggering neurological and psychiatric symptoms. There are many theories as to why. Some studies suggest that gluten has properties that can have opioid-like effects. Others point to the bidirectional nature of the gut-brain-immune axis. We know that in various populations it's been shown that gluten contributes to inflammation, that inflammation can promote gut dysbiosis, and that gut dysbiosis can cause the intestinal hyperpermeability that allows inflammatory molecules to escape and make their way through to the blood-brain barrier. This increases the BBB's permeability, and subsequently these inflammatory molecules infiltrate the brain, affecting mood, including anxiety and depression.

As far back as the 1950s, it was reported that removing gluten from the diet often had a positive effect on the behavior of "naughty" children without celiac disease. In 1966, researchers found a link connecting gluten consumption and schizophrenia. In 2012, a girl who presented with severe psychiatric symptoms saw them completely vanish after being put on a gluten-free diet. Despite these reports, it used to be commonly believed that only people with diagnosed celiac disease could be negatively affected by gluten. Over the last decade or so, however, research has shown that for people with non-celiac gluten sensitivity (NCGS)—who test negative for both celiac disease and wheat allergy, though they do frequently test positive for the antibody to gliadin (one of the main proteins in gluten), and who number six times that of diagnosed celiac disease cases—eliminating gluten can resolve a whole spectrum of symptoms. These symptoms are not limited to intestinal distress, but can also affect the brain, skin, joints, and muscle. NCGS patients frequently report a "foggy mind," which includes diminished cognitive and memory skills, as well as fatigue, anxiety, and depression. What we feed our kids does matter, and the influence of food on children's behavior, psychology, and cognition is much greater than is often appreciated.

Teens Will Be Teens

As the boys grew older and continued to do well, they no longer needed regular follow-up visits with me. But when Luke turned about seventeen, Tonya and Richard brought him in again. While he no longer had the violent emotional outbursts, he remained emotional and sensitive into his teen years. Lately, though, he seemed moodier than normal and prone to unusually acute fear and anxiety. A frank conversation with Luke about his diet and lifestyle revealed that as he'd gotten older, he'd become more lax about exposing himself to gluten and dairy. He didn't eat much of either, but because the symptoms can be delayed and subtle, most people underestimate the impact even small amounts of gluten or dairy can have on a body sensitive to it.

A diet like Luke's doesn't work the same way as, say, a calorie-restricted diet. One over-the-top day of pizza, cake, and ice cream is not going to destroy the positive effects of four months of conscious, healthy eating. Yet small amounts of gluten in someone who is sensitive to it can trigger indigestion, brain fog, aching joints, mood swings, and other symptoms that can last days or even weeks. The dramatic mental and physical turnarounds made possible on a gluten-free diet are only as sustainable as a person's compliance. It's a reality that teens often have to learn for themselves the hard way.

I sympathize. It's a pain to swallow a battery of supplements every day, and it's a pain to carefully read labels and ask annoying questions in restaurants. Kids with true allergies have such violent, frightening reactions that they're motivated to stay the heck away from anything that could possibly trigger a flare. Kids with sensitivities sometimes have to touch the flame once or twice to remind themselves why they need to stay away from the fire.

In the interest of being thorough and following up on previously abnormal labs, we did another round of IgG testing on Luke that revealed a previously undetected sensitivity to almonds. Though he'd not been keeping a strict gluten- and dairy-free diet, he had continued to use almond milk in place of cow's milk for coffee and cereal. With so many other options available such as rice, coconut, or even pea milk, ditching the almond milk was easy for Luke.

It's impossible to say if it was the removal of the almonds from his diet, or strengthening his commitment to staying away from gluten and dairy, but shortly after our visit Luke and his parents reported that his moods had stabilized and he was feeling better. Luke is still sensitive to sounds, but he doesn't seem to mind them if he's making them himself. Today he and his brother, Caleb, are accomplished musicians who have toured in various bands, including one in which they play and write music together. Caleb is completing an engineering degree and Luke will start college in the spring. You'd never guess how far they've come.

Out of the Fog

"Cat."

Hannah Mansouri was lucky to be sitting on her living-room floor. If she'd been standing upright, she might have fallen over. She looked over at the little girl gazing closely at a brightly colored board book.

"What's that you said, Leila?" she asked, trying not to sound too eager.

If Leila noticed her mother's excitement, she didn't let on. "Cat, Mama." She pointed to the big red letters below a picture of a cuddly ginger tabby and sounded out the word. "C-A-T."

Hannah's eyes started to water and she left the room to hunt for a box of tissues. Forty-eight hours earlier, Leila couldn't even talk. Now she was *reading*!

Leila is quirky yet quick with a joke and a smile, but she wasn't always this way. As a young girl, she was angry and anxious, locked so far away in her own world that even her parents couldn't reach her. As it turned out, identifying food sensitivities wasn't the answer to her problems, but knowing they existed at all made it possible to uncover and treat other issues that were compromising her health and mental state.

When Leila was a toddler, Hannah tried every common remedy to get rid of the angry red rashes that suddenly sprouted all over her daughter's torso, but neither pure detergents, lotions, nor oatmeal baths seemed to help. As Leila entered pre-K, Hannah was sure something just wasn't right. Her color didn't seem as vibrant as that of the other children in her neighborhood, children with whom she couldn't seem to connect or play. Hannah couldn't tell if the rejection was because of Leila's delayed speech—she had a hard time forming her words—or the fact that she just couldn't seem to keep up physically. She was tall for her age, but ran slower than other kids and her movements seemed uncoordinated. Then again, it might have been because she wasn't particularly nice to the other kids.

Hannah and her husband, Omar, loved their daughter, but even they could see that Leila was an angry and withdrawn child. They repeatedly brought up their concerns with Leila's pediatrician, a nice woman running a busy practice, but since her speech and physical delays still fell within the normal range for a child her age, the doctor didn't see anything to worry about. Every child comes into their own in their own time, she assured them. Be patient.

Like any parent whose child seems to be suffering or unhappy, Hannah wasn't interested in cultivating patience. Thinking the rash might be making the girl so uncomfortable she couldn't focus on anything else, Hannah made an appointment with an allergist, hopeful that a visit might clear up the mystery. But Leila's test results were negative. No allergies. Hannah felt like everyone around her was touting the benefits of going gluten-free, so she asked the allergist if he thought that might be a good idea. "There's no reason to do that," he replied. "Besides, it's too hard."

So it was a shock when about a month later, an integrative family physician Hannah had sought out said thoughtfully, after spending about ten minutes observing and talking to Leila, "You know, she's extremely bright, but she's just not in the room."

Not in the room? What did that mean?

"She's out of it. And kids like that, I find they often do really well when we get rid of the gluten and casein in their diets."

The doctor urged Hannah to put Leila on a challenge diet. For two weeks, she'd keep Leila away from any trace of gluten and casein, a protein found in milk, and take note of any changes. After two weeks, she would reintroduce gluten and dairy and again observe what happened.

Contrary to what the allergist had told Hannah, going gluten-free, and even dairy-free, wasn't that hard. It was 2008, and the market for gluten- and dairy-free products was exploding. Hannah spoiled Leila rotten, buying every delicious-looking toddler-approved treat she could find from her local health food store and supermarket—cookies, crackers, ice cream, chocolates, chicken nuggets—all of it gluten- and dairy-free. Leila didn't even realize anything was missing.

Leila acted so strangely the first day that Omar wanted to take her back to the doctor. She seemed slow, and her speech, already hard to decipher, was unintelligible. It was late on the second day, however, when Leila suddenly seemed to come out of her fog, picked up a book, and in a clear voice, began to read. A few days after that, Omar's sister came over with her children, and the mothers decided to take the gang to the park. There, Hannah watched, gobsmacked, as Leila sped ahead of her cousins and leaped fearlessly onto the playground equipment. She had clarity. She had courage. She was a different child.

The transformation was so dramatic that when the two weeks were up, Hannah didn't reintroduce the gluten and casein into Leila's diet, though she did cut back on the treats. Six weeks later, the eczema covering Leila's torso was gone. From then on, Hannah usually knew when Leila had eaten something with gluten because almost immediately the rash would come back.

Hannah thought that the story would end there. Unfortunately, it didn't. Removing the gluten and dairy had made a huge difference, but eventually Hannah had to acknowledge that something still seemed a little off. Leila could communicate better, but she still wasn't as articulate as some of her peers. More troubling, she seemed to have difficulty making eye contact. Now that she was in preschool, her anxiety and quick temper were making it hard for her to connect with friends and teachers, and her hyperactivity was becoming problematic. Everyone could tell she was extremely smart, but she still seemed not fully present. In addition, her mood swings were epic.

Around the age of four, Leila was diagnosed with high-functioning Asperger's (which explains her ability to read so young; it's not unusual for kids on the spectrum to be hyperlexic). When she was five, a behavioral optometrist prescribed prism lenses, special glasses that are often used to treat a condition known as "convergence insufficiency," in which the eyes can't coordinate properly, affecting depth perception. For people with this condition, the world can feel precarious and off balance, as though every step might send them hurtling off a cliff.

This could've easily explained Leila's anxiety, as well as contributed to her hypersensitivity and hyperactivity. After a few months wearing the prism glasses, Leila seemed to feel more secure and started interacting better with her peers at school.

The mood swings and tantrums continued, though. On a particularly rough day after school, Hannah found a half-eaten pack of fruit gummies in Leila's lunch box. Though parents had been trading anecdotal stories about how artificial colors and dyes might be connected to hyperactivity and other behaviors in children for decades, right after Leila was born a prestigious medical journal had published the results of a randomized, double-blind, placebo-controlled trial that firmly established a link between food additives and hyperactivity in two cohorts of kids, three-year-old toddlers and eight-to-nine-year-olds. Hannah wasn't aware of the publication, but it probably explains why the idea that some kids might be sensitive to artificial dyes had gained common currency. As she studied the cartoon character grinning maniacally from the front of the fruit gummy package, she wondered if there was a connection between the days her daughter got extra snacks from her school or friends and the days when she couldn't seem to control herself. It was worth finding out.

Hannah arranged with the school that Leila shouldn't be given any food that wasn't brought from home. It wasn't that hard—the school already knew they needed to let Hannah know ahead of time about birthday or holiday celebrations so she could be sure to provide a special cake or treats that Leila could enjoy alongside her classmates. They just needed to make sure no one was sharing any snacks at the lunch table. Sure enough, Hannah and Omar both noticed that Leila seemed to be having fewer bouts of rage. Every time Hannah adjusted Leila's diet, she saw an improvement.

For a few years, Leila seemed to be doing okay, though she perpetually struggled to make close friends. Then she started fourth grade and her parents noticed a major backslide. She was hyper-aggressive, melting down multiple times per week, refusing to do her homework, and barricading herself in her room with furniture, swearing she'd never speak to them again. It really wasn't that uncommon, her teachers assured them. Lots of kids become overwhelmed as the pressures and responsibilities of school ramp up and they realize more is going to be expected of them. She'd get over it.

Months went by and Leila didn't get over it. More than once she ran away, though her parents were always easily able to track her down at a neighbor's house. The whole family was in turmoil. In 2012, Hannah and Omar decided that they should move to a rural area where they hoped Leila could get more outdoor time and fresh air. The move brought them within easy driving distance to me. They decided to bring her to my office to see if there was anything going on that might have been overlooked.

Phenols: The Final Clue

Leila was subdued in my office. In fact, she was completely withdrawn, barely able to bring herself to answer my questions. I did find out that she dreaded going to the bathroom because her stools were so large and hard. There were no signs of the rage and aggressiveness Hannah swore was an almost daily part of their lives, but it seemed it would be a good idea to test for Lyme disease.

I started her on methyl B-12, which I've found can be calming and helpful for children with Asperger's syndrome, a high-functioning condition on the autism spectrum in which the impairments are mainly found in communication and socialization skills. I gave her magnesium taurate to help with the constipation (it was also calming), and since Leila couldn't tolerate the high concentration of probiotics present in capsules, I told Hannah to look for coconut kefir, a fermented beverage similar to yogurt drinks, at the supermarket. Then, when the Lyme disease test came back positive, I started Leila on a round of antibiotics.

In our follow-up visit, Hannah reported that the meltdowns had diminished significantly, and Leila informed me that coconut kefir, which had worked to soften up her stools and make going to the bathroom a nonevent, was awesome. But now I noticed something new: Leila's cheeks were red. Along with hyperactivity, aggression, and a slew of other symptoms that often overlap with other conditions (including Lyme), red cheeks or ears can be a symptom of phenol sensitivity. It made sense to suggest another challenge diet, this time eliminating phenols.

Phenols are a naturally occurring chemical compound found in many foods we would normally encourage our children to eat—fruits like berries, apples, grapes, raisins, and bananas; vegetables like green peppers, tomatoes, and tomato products; and almonds and almond milk. Phenols also appear in artificial colors, flavors, and preservatives. Although most highly colored vegetables are high in phenols, carrots and beets are not, nor are pears. Phenols are also found in corn, wheat, rice, legumes, and even spices, including turmeric.

To quote Hannah, it made a radical difference. "Girl is totally chill," she told me when we next followed up. Bored, because of course no one wants to eat chicken and pears for three weeks straight, yet calm and even tempered. As soon as Hannah tried reintroducing phenols, the intense aggressiveness returned.

Nine months later, Hannah and Omar were astonished at how much Leila had changed. Her moods were even, her temper manageable, and best of all, her Asperger's symptoms were vanishing. It was impossible to know what had made the greatest difference. Was it the methyl B-12? The antibiotics that eradicated the Lyme disease? The phenol-free diet? It was surely all three, but considering how over the course of her whole life Leila's moods seemed to be affected by what she ate, Hannah and Omar suspect much of the credit goes to an improved diet. They know it's anecdotal, but for them the evidence is clear: Milk was Leila's "beast," turning her irritable and mean. Gluten made her a little dopey and fuzzy-headed. And phenols made her intensely aggressive and defiant. Regardless, there's no doubt that Leila's transformation was made possible by slowly lowering multiple levels in her immune kettle.

Growing Pains

It's not uncommon for kids, especially girls, to start feeling intensely anxious as they enter their teen years and begin middle or high school. Leila was no different. A year and a half after her final follow-up visit, her mother brought her back to see me. Leila was now twelve years old, in sixth grade, and Hannah was concerned about the levels of anxiety she could see creeping up in her daughter. Leila started having panic attacks and finding it almost impossible to make presentations in front of her teachers and classmates. There were nights when she couldn't sleep, plagued with fears of failure. She refused to go to summer camp or try anything new. Her relationships with other kids were rocky because she tended to overreact to small slights, real and imagined. It was as though she was always ready for the kind of rejection she'd experienced as a little girl, when she wanted so badly to be a part of the group yet couldn't figure out how to fit in.

Everything I heard sounded perfectly normal for a young teen. In other cases, I might have encouraged the family to take Leila to talk with a child therapist, but Hannah was concerned they might suggest putting Leila on psych meds, which she wanted to avoid. Besides, as Leila grew and entered puberty, it made sense that we'd have to revisit her case. She'd responded so well to dietary modifications and supplements, I thought it was likely we could get her feeling better with just a few tweaks to the regimen we'd laid out a few years earlier.

Treatment

I boosted her dose of magnesium taurate, which had worked well for her, and added a bit more taurine, which can also be calming. In addition, I started her on a tiny dose, 5 milligrams twice a day, of propranolol, an old-style beta blocker usually given as a blood pressure medicine. Some adults use it to combat stage fright, and in children, particularly those with Asperger's, it's been found that it can work well to decrease anxiety. Finally, testing showed her to be a little low in vitamin D and iron, so I started her on supplements for those as well.

The last time I saw Leila, she was thirteen years old and doing fantastic. She was sticking to her gluten-, casein-, phenol-free diet, and her fears and anxieties were markedly diminished. She was doing well in school, and felt more confident and secure with her friends. Sometimes she gets gloomy, but she's noticed it often happens when she forgets to take her vitamin D. Otherwise, her energy and mood tend to be pretty upbeat. She still has Asperger's, and sometimes needs help learning how to cope with her environment—supermarkets, for example, with their bright lights, crowds, and music, feel overwhelming and scary to her—but she can look you right in the eye and hold a conversation just like any other kid her age. Remarkably, she has even discovered that she can eat goat cheese and goat milk yogurt without feeling any negative effects.

Fixing Our Diets: An Overlooked Solution

We're fortunate to live at a time when we have access to an array of increasingly effective anti-inflammatory medicines, in particular a new category of biologics that block the mechanisms that lead to inflammation before they can even start. They're expensive, however, and often have side effects. In addition, in cases like the four children we just met, inflammation doesn't always look like inflammation. It can look like diarrhea, cramps, or pain, but it can also look like Leila's hyperactivity, or Eden's inattentive ADHD, or Luke's aggression and sensory overload, which can delay treatment.

In any event, prescribing medicines still only treats the symptoms, not the root causes. We cannot continue to overlook the impact of food and diet on our health as we become more aware of the prevalence of allergies and sensitivities in our children. I'll often warn my allergic or sensitive patients that symptoms that took a long time to build can take a long time to disappear, and then find myself eating my words when their parents call me two weeks later to report their child is happier and more comfortable than they've ever been now that they're no longer eating gluten, eggs, corn, or milk. I continue to remain astounded at how such a simple change can produce such dramatic, powerful, positive results.

COULD YOUR CHILD HAVE A FOOD ALLERGY OR SENSITIVITY?

Did your child have eczema as an infant or small child, or do they still have it?

Was your tween or teen a cranky, fussy baby and toddler?

Does your child have chronic nasal congestion or asthma?

Does your child ever get red ears or red cheeks after eating?

Does your child complain of abdominal pain, bloating, gas, diarrhea, or constipation?

Does your child note fatigue or brain fog after meals?

Does your child get hyperactive after eating foods colored with artificial dyes?

Does your child seem more aggressive after eating?

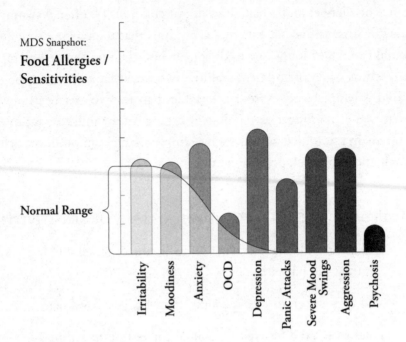

CHAPTER 9

Bottlenecked

*Nutritional Deficiencies, Toxicants,
and Metabolic Disruptions*

A child's position on the Mood Dysregulation Spectrum can be affected by subtle metabolic shifts as easily as by full-blown disease. Every time we put food into the twenty-four-hour energy factories that are our bodies, raw materials in the form of macronutrients (large nutrients, like proteins, fats, and carbohydrates) move down production lines to be identified and funneled to their proper departments, where they undergo a chemical reaction that either breaks them down into usable energy or prepares them to help conduct other functions like growing cells, contracting a muscle, or even breathing. The catalysts that speed up those transformative chemical reactions are called enzymes.

Like a snap-fit joint on a plastic toy, every enzyme is designed to bond with one particular molecule, called a substrate. But just as a factory's output is negatively affected when assembly line workers come in sick and slow, metabolic production is hampered when enzymes show up for work in bad shape—literally. If a child receives a gene mutation that changes the shape of an enzyme or affects its ability to bond to its substrate (the stuff going in), it simply can't process the substrate into a product the body can effectively use.

In some cases, the production line gets jammed, ceasing output altogether and simultaneously creating a heavy buildup in front of the blockage point. In other cases, the machinery might be functional, but not work as well as it should. Production slows down; quality suffers. Skin that doesn't receive enough moisture might develop a rash, insufficient dietary fiber in the digestive tract causes constipation. For some kids, an excess nutrient accumulation or, more commonly, nutritional deficiency, or the combination of both, can have an outsize effect not just on the efficacy of their bodily functions, but also on their mood and mental state, leading to neurological or psychological symptoms.

Nutritional Deficiencies

You've probably heard of some of the diseases caused by nutritional deficiencies. Just think of sailors who traveled long distances on seafaring ships suffering from a lack of vitamin C. The resulting condition, scurvy, prevents the formation of collagen, a key component of connective tissue, which explains why the disease can ravage multiple body organs. There's also rickets, which deforms the bones, as a result of a diet low in vitamin D and/or calcium. What's less known is that these and other nutritional deficiencies can also have psychological symptoms. Low magnesium levels may lead to low mood and even depression. Blocked metabolic processes related to deficiencies in zinc and B vitamins, such as niacin and thiamine, can manifest as neuropsychiatric symptoms like anxiety and OCD.

I wish the majority of my patients' nutritional deficiency diagnoses were cut and dry and that I could heal them with a single medicine or supplement. Pellagra, for example, which can cause the 4-Ds—diarrhea, dermatitis, dementia, and even death—can be reversed with a supplement or injection of vitamin B-3 (niacin); pernicious anemia can be resolved in a matter of weeks with vitamin

B-12. Still, as we'll discuss, it's rare for any one micronutrient (small nutrients, like vitamins and minerals) to make a difference all by itself. Getting most of my patients to optimal health is generally a process of first clearing away what's harmful, then layering and combining treatments to see what's most helpful. Introduce too many nutritional supplements too early or in the wrong sequence and you might see no effects, but that same treatment might work beautifully when delivered in the right doses in the right order. The conditions have to be right for the body to be able to make the best use of nutritional supplements, usually in combination with other nutrients or even pharmaceuticals.

One of the most common nutritional deficiencies I see is methylfolate, especially in my patients with mood disorders. Methylfolate is necessary for building neurotransmitters like serotonin, dopamine, and norepinephrine, which are responsible for promoting feelings of happiness and emotional well-being. If you don't have sufficient methylfolate, you can't make these neurotransmitters, which can leave the brain primed for depression and anxiety.

Methylfolate is the active form of folate (the same vitamin pregnant women are encouraged to take to ensure healthy fetal development and prevent neural tube defects), created during a metabolic process known as methylation. It's important to understand that even if you have sufficient folate in the body, you may not have enough methylfolate in the brain—because methylfolate is the only form of folate that can cross the blood-brain barrier. In fact, one study showed that up to 70 percent of patients with depression have tested positive for a gene mutation that compromises the body's ability to metabolize folate into methylfolate. Methylfolate supplements have been shown to be such an effective treatment for depression that the FDA has approved them for patients who can't tolerate or who show an inadequate response to antidepressants.

Emily was one of those patients. At fourteen years old, she didn't want to take the multiple supplements she needed to help her keep her body and mind in optimal shape. But no matter how rebellious or resistant she felt, Emily always took the methylfolate because she remembered all too well how dark her world felt without it.

Emily's parents, Madeline and Travis, often traveled to Central America on mission trips with their church. When they spotted baby Emily in an orphanage it was love at first sight, and by her eight-month birthday she was the newest member of their family, joining their six-year-old biological son, TJ, and their three-year-old daughter, Corrine, who was born with cerebral palsy. Emily's little body showed some signs of neglect—her bottom was lightly scarred from constant diaper rash—but she was a smiling, happy baby. However, Madeline and Travis knew that any trauma, even trauma that occurred long before a child could form memories, could lead their new baby to have some attachment difficulties as she got older. From day one they consulted regularly with a family therapist specializing in adoption-related issues. Armed with boundless love and professional guidance, the family thrived.

Then, around age seven, Emily started throwing multiple tantrums per day. She'd been a strong-willed toddler, so it took the girl punching a hole in the wall and other over-the-top behaviors for Madeline to realize that she was witnessing something unusual from her child. She took Emily to her pediatrician. After listening to Madeline's description of Emily's behavior, he took Madeline by surprise by suggesting a throat culture. A throat culture? Why, when Emily wasn't sick? Madeline was shocked when the test came back positive for strep, and even more shocked when just a few days after taking the prescribed antibiotic, the temper tantrums disappeared and the old cheery (though still strong-willed) Emily returned. Unfortunately, two or three weeks later Emily's destructive behaviors returned. Once again, a throat swab revealed strep, and once again, the tantrums disappeared after a week of antibiotics. The next time Emily started throwing things, the doctor didn't even bother with a throat swab. He extended the antibiotics for thirty days. After a few more rounds of this cycle, he diagnosed Emily with PANDAS.

This didn't sit well with Madeline because Emily's blood tests never showed elevated antistreptolysin O antibodies (ASO titer), which would have supported the presence of strep (although a positive throat culture is the most compelling evidence for the presence of strep). Still, there was no doubt the antibiotics worked, albeit temporarily, so she went along with the diagnosis. After two years, however, she decided to find a different long-term solution. Emily was constantly complaining of abdominal discomfort and painful constipation; Madeline was sure the antibiotics were destroying Emily's gut.

She was also worried Emily's behavior would destroy their family. Terrified the neighbors would hear Emily's screams and call Child Protective Services, Travis and Madeline sold their house and moved to a different subdivision where there was more distance between properties. They also couldn't keep forcing Corrine to share a room with her sister. Emily had smashed her fist through the bedroom window, and the girls' walls were pockmarked with holes. In the new house, Corrine had taken to hiding in her bedroom with the family dog, who would start to shake as soon as he heard Emily's voice rise. TJ immersed himself in sports and after-school activities, staying out of the house as much as possible.

As soon as she'd seen Emily's behavior worsening, Madeline had connected her daughter with a therapist, but the counselor's words weren't reassuring. "Be prepared for Emily to be diagnosed with bipolar disorder as an adult." (In fact, symptoms of bipolar disorder do occur in children and can be diagnosed at an early age.) A renowned child psychiatrist echoed that prediction as well, and prescribed an SSRI, which made Emily ten times worse. He switched her to another SSRI and that, too, increased the frequency and intensity of her meltdowns. The way he spoke to her, Madeline could tell the doctor was preparing her to consider antipsychotics or even institutionalization. But she wasn't willing to take that step until she had exhausted every other option. She knew she had to act quickly, though. Emily, now ten years old, had started describing herself as feeling "black inside." In family photos, Emily's smile seemed fake.

Madeline and Travis were both civil servants, and money was tight with two other children, one with her own medical issues. They'd put everything they had into making sure Corrine received the best care and every possible treatment available to her, with so much success that by the time she was a teenager, at first glance you'd never know she had a motor disability. Their second daughter deserved nothing less. Madeline and Travis had read about me and wanted to bring Emily to see me, but they also knew that insurance might not cover many of the tests they expected I would run. They made an appointment anyway. When they showed up in my office with Emily, they had a commitment from Travis's mother to help with any uncovered medically related expenses.

After I listened to their story, Madeline remembers me saying, "Take heart."

"I'll take whatever you can give," she thought.

Something jumped out at me the first time I looked at Emily's medical records. Emily had a double MTHFR gene mutation, meaning she had received one mutated MTHFR gene from each parent. A double mutation in this gene (homozygous C677T) makes it 60 to 70 percent less efficient at its job, which is to help create the enzyme necessary to convert folic acid into methylfolate. Methylfolate is critical, enabling a host of processes including cell regeneration and DNA repair, and as we mentioned earlier, creating the neurotransmitters serotonin, dopamine, and norepinephrine, as well as the hormone melatonin, which we need for proper sleep. There are several variations of the mutation, and Emily's sister, Corrine, had one that was the suspected cause of an in utero stroke that led to her cerebral palsy. Just to avoid leaving any stone unturned, Madeline had asked the pediatrician to test for the mutation. The positive result was what had cinched her and Travis's conviction that they needed to get Emily in front of someone like me.

Emily had no memory of the months she spent neglected in an orphanage, yet multiple studies note that children who experience early trauma often show a predisposition to mental health issues like depression later in life. Add in the double MTHFR mutation that compromised her ability to produce methylfolate, and she, like many with the MTHFR mutation, was a prime candidate for a mood disorder, because anything that hinders the methylation process like an MTHFR mutation is also hindering the body's ability to cope with stress. In addition to creating neurotransmitters, methylation enables the production of glutathione, a key antioxidant, detoxifier, and free-radical scavenger in the brain. For kids with Emily's kind of genetic vulnerability, it's like the biological cleanup crews that normally rush in when the body is confronted with emotional or environmental stressors are trapped behind locked doors. Without it, the toxins and stressors simply build up and cause greater damage.

I tested Emily for mycoplasma, Lyme, and other infectious agents that might cause a PANS-type response (what I prefer to call ITABI, as you might remember from chapter 4), but they all came back negative. Her metabolic and nutritional workup, however, showed vitamin D deficiency. Low vitamin D is often found in patients with inflammation, and it's a common finding in kids with ITABI. In addition, vitamin D seems to improve symptoms in children with tics. Emily didn't have tics, but many children suffering from ITABI do. Vitamin D also helps protect the integrity of the blood-brain barrier (BBB). You'll recall what can happen when certain stressors cause the tight junctions of the BBB to loosen—it becomes more permeable, therefore making it possible for large inflammatory molecules to pass through and wreak havoc.

Consistent with the MTHFR mutation, Emily also tested low for glutathione and sulfate. Glutathione is formed at the end of the methylation process, so when there's a hiccup or defect like, say, an MTHFR genetic mutation or something else that blocks the production of methylfolate, its production is compromised. Like glutathione, sulfate is also important for detoxification. In fact, the sulfation cycle plays a role in forming glutathione, so when one is low, the other often is, too. Low levels of glutathione likely mean higher levels of neuronal toxicity.

Emily was born genetically vulnerable to the effects of stressors.

My suspicion was that the combination of the strep infections and the antibiotics Emily had taken as a small child launched a chain reaction of cellular stressors that compromised all the protections her body had in place, weakening her gut lining and her brain-blood barrier. While I could not identify an infectious trigger usually associated with ITABI, other signs pointed to brain inflammation that was manifesting itself as psychotic behavior and depression.

Treatment

Successful treatment for Emily's low methylfolate would depend on working around her genetic mutation. By giving Emily methylfolate in supplement form, we could give her body what it needed to start producing the neurotransmitters necessary for mood regulation and the glutathione necessary for detoxification.

I also started Emily on an injection of methyl B-12, which is the form of vitamin B-12 that works best on the neurological system and brain. It's not just a methylator; at times it can be a miracle worker. Methylators seem to work better when offered in combination, and methylfolate and methyl B-12 seem to work especially well together. They can increase glutathione. I've seen autistic children speak for the first time after taking them. It's also been used in treatments for chronic fatigue syndrome, Alzheimer's disease, and neurological aging.

Vitamin D deficiency has been linked to depression, possibly because it causes an increase in inflammatory factors that can compromise the blood-brain barrier. Vitamin D supplementation has been shown to mitigate moderate to severe depression in patients over the age of sixty, as well as depression in teenagers. I gave Emily a therapeutic dose of vitamin D to help restore the integrity of her blood-brain barrier, modulate her immune system, and decrease inflammation. I also prescribed low doses of magnesium (too high a dosage caused loose stools), iodine, and vitamin E, the latter of which I like to give alongside omega-3 essential fatty acids to protect them from oxidation and to enhance their anti-inflammatory effect.

Next, I addressed Emily's diet. At first I simply advised her to avoid food additives and refined carbohydrates to get the sugar out of her diet, as well as limit her dairy intake, but eventually I urged her to also go gluten- and dairy-free to help minimize neuroinflammation. To compensate for the lack of dairy in her diet, I had her take extra calcium. In addition, I started her on fish oil, curcumin, and other nutrients and herbs with well-documented anti-inflammatory properties to help decrease the inflammation in her brain.

And of course we needed to help kill the bad players in Emily's gut caused by the years she spent on antibiotics, and restore the gut lining. A round of nystatin, an antifungal medicine, would also help resolve the gut dysbiosis, while probiotics and prebiotics would help repopulate her body with good bacteria.

At the first follow-up, Madeline and Travis reported slow but gradual improvement. While some symptoms were abating, they'd noticed an uptick in Emily's hyperactivity. I adjusted her methylating nutrients downward and added hydroxy B-12 to the methyl B-12 injection, which can work well for kids who get hyper or agitated with methyl B-12 alone or with methylfolate. The needles are tiny and cause almost no pain, but many kids simply can't handle the anticipation of getting injections. Emily was one of them. After about a month, she refused to accept them anymore, so I switched her to a nasal inhalation spray, which she also disliked but was willing to use even though the methylcobalamin component is red and can leave traces around the nose that make it look like you have a nosebleed (so be forewarned!).

The change, in Madeline's words, was huge. Their daughter was less argumentative and anxious. Her constipation had resolved and her abdominal pain was gone. She was even-keeled and able to self-regulate. And when I saw them a year later, Madeline announced, "She's healed." Emily concurred. "I feel like myself," she told me. Her broad smile reached her eyes. No faking there.

It's important to Madeline to acknowledge how lucky the family was to have Travis's mother as financial backup through this whole ordeal. "We rely so heavily on private insurance, but then they don't pay for the things that are really going to help you. They don't pay for testing, they don't pay for the B-12 or the supplements. But they would have paid for all of Emily's antipsychotic medicines if we had gone that route. It's crazy." Madeline received one more shock about a year later. In 2018, Emily's therapist discharged her, declaring her a well-adjusted, happy adolescent. "Those were words we never thought we'd hear," says Madeline.

Emily's case is a dramatic example of what can happen with a serious methylfolate deficiency, but methylfolate can also make a huge difference in the life of kids who just need a little help making their way through those rough teenage years. In some instances, it can also allow kids who take antidepressants to lower their dosage and may possibly even eliminate the need for medication altogether.

Adjusting as Kids Grow

Jennifer's story begins when she was a child. Her mother, Diane, described her daughter as a bubbly little girl who would suddenly and regularly dissolve into a lethargic puddle. Purple circles would form under her eyes. She'd eat and complain of a stomachache. Her frequent head colds turned into respiratory infections, and she landed in the hospital with pneumonia at least twice. The doctors would put her on an antibiotic and she'd get even sicker until steroids were added. Running and playing would frequently trigger an asthma attack. After several years, when Jennifer was about seven or eight years old, her pediatrician wasn't sure what else to do, and recommended Diane take her daughter to see me.

After all those years of antibiotics, I was sure Jennifer had candida—also known as fungal dysbiosis or, more colloquially, yeast in the gut—and that it was downgrading her immune system. My first plan of action was to put her on a gluten-free, dairy-free, yeast-free, sugar-free diet and start a regimen of probiotics. I also prescribed transfer factors, which along with everything else would help modulate Jennifer's immune system.

Within the first year Diane saw a remarkable transformation. Her daughter was happy and healthy. No more need for nebulizer treatments, no more antibiotics. Keeping her daughter on a strict diet, however, was difficult. This was the early 2000s, when gluten- and dairy-free products were still considered fringe and hippie food. The options were few and mostly terrible. Aside from meat and vegetables, Diane may as well have been feeding the child tree bark and sawdust. It took a ton of research and time in the kitchen to put together something palatable, but the effort paid off. On her strict diet and nutritional and herbal supplements, Jennifer went several years without so much as a cold.

Even at such a young age, Jennifer recognized that the diet and supplements were having an incredible effect on her energy and quality of life. She loved never missing a day of school and she was relieved to steer clear of the doctor's office. Jennifer hadn't even realized how much anxiety she was carrying all the time until it was gone and the world looked brighter.

But even though she was happy to feel better, no little kid likes to feel different from their peers. No one wants to have to explain why they brought blue corn tortillas topped with a little homemade tomato sauce to the pizza party. No one wants to hear how "weird" it is that they don't eat pizza or ice cream, despite knowing the comment isn't intended to be malicious. Those words stung anyway.

The second part of Jennifer's story begins not too long after she started her menstrual period. As she'd grown into adolescence, Jennifer had shown a tendency toward depression, but Diane also knew that girls could get moody and sad during these rough years when peer pressure increases and their self-esteem takes daily hits. Like many young teens, Jennifer's moodiness tended to spike monthly, right around the time her period would start. Also like many young teens, her flow was heavy, and painful cramping was a regular problem that frequently resulted in phone calls for Diane to come pick her daughter up from school.

As Jennifer started high school, Diane began to notice that her daughter seemed to be withdrawing. A perfectionist with a studious nature, Jennifer had always been partial to artistic or solitary activities like painting and reading, but now she was disappearing into her own little world. When she did join family or friends, the exchanges were emotional and volatile with exaggerated highs and lows. Again, at first Diane didn't think too much of it because teenagers and emotional volatility go hand in hand. But the lows were getting more frequent and drawn out, and the pressure Jennifer put on herself was intensifying. No matter what Jennifer achieved—a prize in a student art show, a straight-A report card—all she could see was what wasn't going well and where she believed she was failing. She wanted to be valedictorian, and nothing else would do. Diane knew this wasn't ordinary teenage angst and ambition. What she didn't know was that Jennifer had become lax about her eating habits when she wasn't being supervised at home.

When Jennifer's gynecologist put her on birth control pills to help ease her painful monthly symptoms, Jennifer went off the rails. Her depression deepened, manifesting as massive mood swings with prolonged crying jags. Diane thought hormonal contraceptives were supposed to ease PMS and cramps, and worried about how much worse the symptoms would be if Jennifer wasn't on the birth control pills.

I still saw Jennifer regularly for checkups. When I heard what was going on, I suspected she was having a negative reaction to the birth control, so I suggested she talk to her gynecologist about switching her to something else. A hormone-releasing IUD made no difference. Jennifer only found relief from her intense menstrual cramps and mood swings after starting an ultra-low dose of a birth control pill. But considering her history even before she went on the pill, I was convinced she was struggling with underlying depression, and started her on the antidepressant Lexapro.

Jennifer immediately felt the medication helped, but she also started experiencing awful headaches and gaining a significant amount of weight, which was the last thing a girl struggling to fit in and feel accepted by her peers needed. The side effects were so bad she refused to keep taking the Lexapro. A different antidepressant caused the same discomfort and weight gain. Jennifer was turning on herself more and more. Diane was distraught.

Treatment

I'd hoped the antidepressants would keep Jennifer from falling into her deep emotional lows while simultaneously boosting her mood stability and energy through diet and supplements. Now we were going to have to help her regain balance through diet and nutrition alone, a tougher proposition, but not impossible. Over time, guided by labs and Jennifer's clinical response, we would add multiple supplements to Jennifer's diet as we figured out what worked best for her. I pinned my hopes on three supplements in particular.

1. Methylfolate: Jennifer never knew instability or neglect the same way that Emily did, but the inability to breathe is in and of itself pretty traumatic, as is worrying about the next unexpected respiratory illness or asthma attack with accompanying doctor visits. It's possible that she, too, had a predisposition to depression caused by her early childhood experiences. In addition, even without testing positive for an MTHFR mutation, Jennifer's history of low-lying depression and perfectionistic tendencies is typical of patients who can't properly process folate into a usable form. I prescribed a methylfolate supplement in addition to another supplement called S-adenosyl methionine (SAMe), another key player in the methylation process that ultimately creates neurotransmitters. SAMe has been shown to augment positive effects when combined with methylfolate or other treatments for depression, helping to balance patients' dopamine, serotonin, and norepinephrine levels for proper mood regulation.

2. Vitamin D: Like Emily, Jennifer's testing revealed a Vitamin D deficiency. Now, most Americans have insufficient Vitamin D levels—
   possibly due to a confluence of factors, including lack of adequate sun exposure (especially for those of us who live in northern latitudes where it can be overcast for months), increased use
   of sunscreens and cover-up clothing, and insufficient vitamin D
   from food sources—and most people don't experience any obvious
   ill effects. But a patient like this—who'd already sustained so many
   assaults on her immune system and who very likely had come
   across a pathogen in her early life that had weakened her immune
   system and blood-brain barrier—needed every advantage to help
   her regain her health. Optimal levels of vitamin D would be an
   important weapon in her battle to regain her physical and mental
   health.

3. Glutathione: A urinary organic acid test (which can detail excess
   or low metabolic products in our system better than a blood
   test) suggested Emily had low glutathione, which would make
   sense. A body that wasn't methylating properly would also have a
   hard time producing adequate glutathione, a key member of the
   cleanup crew necessary to rid itself of free radicals and environmental toxicants like heavy metals.

These in addition to other supplements and a course of an antifungal medication made all the difference for Jennifer. Within a few
months, Diane could see a notable difference in her daughter's
behavior and even her body language.

Like most kids, once she didn't feel sick anymore, Jennifer
stopped being as careful about her diet, sometimes intentionally,
but often as a result of the spontaneity and ambivalence that's part
and parcel of the teenage years. Though she became more lax, Jennifer was still acutely aware that she probably needed to be more
careful than most. While on one hand this awareness kept her
from experimenting with drugs and alcohol to the same extent as

some of her classmates, on the other hand it served to make her feel isolated from them. Diane doesn't think Jennifer really began to believe in herself until she got an acceptance letter from her first choice university. In the years after, Jennifer succeeded spectacularly and is now working as an architect on the West Coast. She thinks she'll always be prone to depression, but so far she's been able to manage her mood by sticking to a healthy diet and taking her supplements. She also credits a regular yoga practice and talk therapy with helping her maintain her mental and physical strength.

Fighting Fear

While Jennifer's anxiety manifested as intense pressure to succeed and perfectionistic tendencies, Eric's looked a little different. He was a hoot with his mischievous smile and a white-blond tuft of hair that reminded me of a baby chick. Eric was a well-adjusted eight-year-old who got along with his parents, had friends, and did well in school, but he would go into a panic when he found himself in situations he couldn't control. Anytime they traveled by air his parents gave him a dose of a strong antihistamine to calm him down and make him drowsy. It wasn't that Eric had a fear of flying, he just became extremely anxious at the prospect of any uncertainty or confusion once they got off the plane. The anticipation of a tight flight transfer or questions about where they'd catch a ride to their hotel could twist him into a taught, panicky wire. He also had an intense oral fixation so he constantly had something in his mouth, usually his shirt, a pencil, or his fingers.

Eric was seeing a very good doctor who had discovered the boy (like Emily) had the double MTHFR genetic mutation, and had prescribed a sublingual combination of methyl B-12 and methylfolate. Unfortunately, it didn't seem to help, so Eric's mother brought him to me. As soon as I met him, I suspected a nutritional deficiency. All the physical signs were there: low muscle tone, significantly smaller than his peers, clinically underweight, and close to the bottom of the growth chart. Sure enough, when the labs came back, he may as well have had no vitamin D in his system. Subsequent testing would reveal that in addition to the MTHFR mutation, he also had a mutation in the VDR gene. VDR stands for vitamin D receptor, another gene that normally allows the body to process vitamin D. Eric was getting next to no vitamin D at all!

Treatment

Generally, methyl B-12 is calming, but we're learning that many people treated for MTHFR can become overmethylated, meaning the supplements are pumping more methyl groups into their body than their system can handle. Among others, classic symptoms of overmethylation include heightened anxiety and panic attacks. This is why I'm extremely careful to use what could be called a Goldilocks method, gradually introducing methylfolate and methyl B-12 until we can confirm the ideal doses that work for each patient.

In Eric's case, one dose per week was too little, but one dose every third day was just right. His diet was already gluten-free because of diagnosed celiac disease, and he didn't get much dairy because both of his parents were sensitive to it, so we just focused on limiting his sugar intake a little more aggressively. In addition to the vitamin D supplement we also added resveratrol—the heart-healthy compound found in grapes, berries, and red wine—which has been shown to restore the integrity of the blood-brain barrier and help protect against inflammation.

Between the ages of eight and ten, Eric grew seven inches and gained twenty pounds. Best of all, he started coming out of his shell. His parents didn't have to manage him the same way when presenting him with new situations. He could fly without being medicated, and played a violin solo for the school orchestra performance. Somehow, the world just didn't seem as scary anymore.

Toxicants

As we've seen, when nutritional deficiencies and gene mutations muck up our metabolism, they can also muck up our ability to detoxify our bodies by depriving us of enough glutathione, which I've referred to as the cellular cleanup crew. Without this antioxidant and detoxifier, it's more difficult for us to flush toxins and pollutants out of our tissues and organs.

Many parents assume that the only children at risk of consuming dangerous levels of toxins are those exposed to lead in pipes or paint, those who live near highly polluted areas, or those whose parents work in certain fields like demolition work or plumbing and can expose their kids to "take home" lead. But the reality is exposure to toxins and pollutants is hard to avoid. In 2016, the World Health Organization estimated that 24 percent of the "global burden of disease," including mental, behavioral, and neurological disorders, is due to modifiable environmental factors.

While we've made great progress lessening the amount of second-hand smoke our children breathe, their environment can still be saturated with heavy metals—from cadmium in their toys, persistent organic pollutants like PBDEs (flame retardants) in their sleepwear, BPAs in their plastic bottles and tableware, phthalates in their soaps and shampoos to PCBs in many of their homes and schools. Thanks to better environmental regulations and technology, atmospheric mercury levels in the U.S. have been falling since the 1970s, with a whopping 30 percent decline from 1995 to 2010. And you'll be pleasantly surprised to learn that it's actually easy to avoid flame retardants in sleepwear, as most clothing manufacturers now make sleepwear with polyester, which is naturally flame resistant. Clothing manufacturers are required to make all sleeping garments for children size 9 months to size 14 flame resistant and able to "self-extinguish," achieved through a tight fit and by making sure the clothes can pass flammability tests. Cotton is flammable, but the requirements for a tight fit make cotton sleepwear flame resistant because it doesn't allow enough oxygen between the child's body and the clothing to allow a fire to "breathe."

And yet as rapidly developing countries increase their emissions, nearly all of our fish supply remains contaminated with the neurotoxicant methylmercury, which is produced by microbes from the inorganic mercury that lands in our water systems. The FDA found mercury in 100 percent of tested samples of canned tuna, as well as in canned mushrooms, shrimp, and crisped rice cereal. For decades, mercury has been linked to irritability, depression, and fatigue, among other neurological symptoms.

Low levels of lead and cadmium exposure are known to damage the brain and have been implicated in playing a role in psychiatric disorders. Their blood levels might not reveal toxicity, but it should be no surprise that some kids with genetic vulnerabilities and overburdened immune systems react strongly if the toxins stored in their tissues (which all of us have) somehow get unbound. There's evidence that prolonged, low-level lead exposure, alone or especially when combined with synergistic reactions between other toxic metals, certain chemicals, and stress, can lead to cognitive dysfunction in the form of decreased IQ, anxiety, or emotional instability. Exposure to lead in early childhood has been linked to lower IQs, "conduct disorder, juvenile delinquency, drug use," and other undesirable outcomes. Even low levels of exposure "are associated with poorer school performance, marked by shortening of attention span, reading problems, attention deficit-hyperactivity disorder, and school failure." Children exposed to lead often show challenging behavior in school, such as aggression, fighting, or rule breaking, but teachers have also reported them to be withdrawn. At least the children displaying externalizing behavior might get some attention and help; the quiet ones, the ones possibly suffering from mood and anxiety disorders, are often completely overlooked. Many older homes still have traces of lead present. As babies become mobile and prone to put anything and everything in their mouths, parents should be aware of the hazards of chipped walls, lead-contaminated dust, even lead-contaminated houseplant dirt.

These are just the toxins in the form of heavy metals and chemicals. Electromagnetic frequencies from computers and cell phones, the stress that accompanies social media, and our culture's increased demands and expectations of children can all, in some way, be considered toxicants as well.

Mold

With climate change dousing us with more frequent, fiercer storms and subsequent flooding and water damage, one toxin getting a lot more attention is mold, which can have a dramatic impact on metabolism, cellular function, and mood disorders. It can affect multiple systems, causing inflammation and mitochondrial dysfunction, and cause myriad symptoms like fatigue, brain fog, headaches, myalgias, fibromyalgia, and neuropathy, to name a few. The primary risk factor for mold poisoning is poverty, which increases one's risk of living in a home with leaky pipes or poor ventilation. But as Barbara Tannen will tell you, mold can lurk in anyone's home without their knowledge.

Until 2016, Barbara and her husband, Trevor, were raising their two children, Katherine and Josh, in a bucolic suburb, spending as much time as possible at their secondary residence, a large rural farm. But when a job opportunity arose, they settled into a beautiful townhouse in a nearby city. Katherine, age twelve, was nervous but excited about starting middle school. Within a few months, however, she started exhibiting some odd behaviors, like insisting on lining up her toiletries in a perfect line before going to bed, and then expressing irrational fears about the safety of her room. She acquired a tic, a quick twitch in her arm followed by a light jerk of the head. She became anxious, sad, and moody. Barbara couldn't figure out if this was a result of self-consciousness due to the tic or the ordinary vulnerability many girls feel in their first year of middle school. The family quickly arranged for Katherine to see a therapist with a specialty in OCD.

After about a year of cognitive behavior therapy, the OCD eased up, as did the tic. But Katherine still complained of constant exhaustion, and still seemed depressed. Her grades were excellent, but she never brought friends over or wanted to make plans with other kids the way she had in the country. Barbara felt like her daughter was paler than usual, too. Barbara didn't want to overreact, so she refrained from taking Katherine to the doctor—what seventh grader isn't moody and tired, right? But then Katherine abruptly dropped out of track, a sport she'd once loved, because she said she couldn't keep up anymore. Now Barbara knew for sure something was very wrong.

School let out and Katherine spent six weeks at a rural summer camp. Barbara was relieved to see pictures of her daughter laughing and playing with other kids. Maybe she'd just needed a break? When Katherine got back, she surprised Barbara by announcing that she wanted to play soccer in the fall. Barbara booked a doctor's appointment so her daughter could obtain the physical necessary for athletic participation, but was only able to secure one toward the end of the summer.

By the time Barbara brought Katherine to the doctor, Katherine was complaining of fatigue again, and the warm, sunny glow she'd carried at camp had faded, along with her smile. The pediatrician suggested they test for Lyme and celiac disease. Barbara understood that their previous life in the country could increase the likelihood of Lyme, but Katherine had never had any notable digestive troubles or abdominal pain. Celiac disease seemed like a stretch, but Barbara went ahead and agreed to the testing. The lab results showed such high levels of antibodies for celiac that the specialist they eventually saw assured her the required endoscopy to diagnose the disease was just a formality; there was no way Katherine didn't have celiac. The biopsy confirmed marked atrophy of the villi. This means the lining of Katherine's intestines, which should've been a mountain range of crypts and peaks, was instead almost flat. In spite of these intestinal structural abnormalities, Katherine had a subset of celiac disease with mainly extraintestinal symptoms, meaning her symptoms manifested outside the GI tract.

Barbara and Trevor got to work purging the kitchen, fridge, and pantry of anything that might have traces of gluten, down to replacing the pots and pans and buying a new toaster. They were elated to finally have the answer to healing their daughter. It was early September, and the doctor assured them Katherine would feel like a new person in a few weeks.

In October, Barbara brought her back to the doctor. Katherine didn't feel like a new person, she felt exactly the same, except now she had a low-grade fever. Repeat testing showed zero trace of gluten, meaning there were no longer elevated levels of celiac antibodies in her system.

The doctor had recently attended a conference on PANS and PANDAS. Katherine hadn't had strep since she was a tiny girl, and she wasn't having massive meltdowns or psychotic episodes, but she did have abrupt onset OCD and three associated symptoms for PANS—tics, depression, and anxiety. Katherine tested positive for mycoplasma, a bacterial infection which has been known to trigger brain inflammation, which can be related to abrupt onset OCD and tics. The doctor thought this might be at the root of all of Katherine's problems. Barbara's research led her to meet other parents of children affected by PANS/PANDAS. That's how she got my name and made her way to my office in Upstate New York.

When I got back my battery of Katherine's tests—which I ran to see what exactly we might be dealing with—my eyes almost popped out of my head. She tested positive for virtually everything: Hashimoto's, PANS, mycoplasma infection, Bartonella, active Lyme, active mono, *and* a double MTHFR mutation. On top of that, her Cunningham panel revealed some of the highest markers for autoimmunity and inflammation of the basal ganglia our office had ever seen. How was this girl even functioning?

Treatment, Part I

The first thing we had to do was knock out the inflammation, infections, and tick-borne diseases. I started Katherine on two antibiotics, doxycycline and rifampin, but her depression deepened until we switched the rifampin to Bactrim. I prescribed prophylactic nystatin to prevent the fungal dysbiosis that can accompany antibiotic treatment as well as started her on a vitamin D supplement to address her vitamin D deficiency.

We also did an IV glutathione treatment. Here, too, we were compensating for Katherine's diminished ability to methylate and produce this important antioxidant and detoxifier. I added intravenous immunoglobulin (IVIG) therapy, because as we've seen, it can work wonders at quelling inflammation in the brain. Finally, I also prescribed her levothyroxine (T4) to treat the Hashimoto's and hypothyroidism.

A month later, we saw slower improvement than I would've liked, but that wasn't surprising given the adjustments we'd had to make in Katherine's antibiotic treatment and the fact that she'd been suffering from a complex constellation of symptoms for such a long time. Every follow-up gives me an opportunity to ask more questions. Trying to think of possibilities we hadn't yet considered, I asked Barbara whether Katherine could have been exposed to mold. She froze. When touring what would become their new home, she and Trevor had noted the basement was damp. The landlord wouldn't do a mold test, but since they didn't see or smell any mold on the walls, they decided that wasn't a deal breaker. And the year before, a terrace outside Katherine's bedroom window had sprung a leak that allowed water to pool in the ceiling of her parent's bathroom one story below.

Barbara hired a professional environmental inspection agency to evaluate the house a week or so later. She was handed a 100-page report on mold, showing the house harbored three types. With all the air units starting in the basement, the ducts had spewed mold into every corner on every floor of the house, even Katherine's, located floors above. Before leaving, the inspector had urged Barbara not to let anyone go back into the house.

The family left that night, sacrificing their security deposit and the remaining three months' rent due on their lease, and moved back to their former home. Barbara paid to have some furniture that could be salvaged professionally cleaned, but threw away every pillow, blanket, stuffed animal, or item made of wicker or wood, whose texture made it easy to hide mold spores even after a thorough cleaning. She bagged five loads of clothes, and every night the family would wash what they wanted to wear the next day with some vinegar mixed into the load. She was so paranoid, after a few months she paid for a mold inspection in their old house. It was clean—except for two spores of mold in the laundry room, stowaways that must have traveled with them in the contaminated laundry.

We tested the whole family for evidence of mold exposure, but only Katherine and her little brother Josh came up positive. Josh seemed unscathed, I suspect due to genetics and other factors in the immune kettle; people seem to have a variable response to the negative effects of mold toxicity. It's likely that at some point Katherine's immune system had been weakened—maybe by the MTHFR mutation, maybe the mycoplasma, maybe the Lyme, Bartonella, or Epstein-Barr virus, maybe a combination of all five—making it unable to protect her against the toxic assault of mold that began when the Tannens moved into their new home.

Treatment, Part II

We progressed to the next level of treatment, which included IV infusions, initially consisting of nutrients including magnesium, vitamin C, glutathione, and phosphatidylcholine, which can be helpful for mold detoxification. Then we added a second IVIG treatment to quiet her brain inflammation. For adrenal support, I gave her oral nutrients and herbal supplements. Due to the inadequate response to the levothyroxine and low-normal T3, I added Nature-Throid, a natural desiccated thyroid replacement. As we discussed in chapter 6, desiccated thyroid preparations deliver T3 as well as T4. Soon after, Katherine's brain fog dissipated and her energy level increased.

Finally, I put her on a mold detox program. I continued the antifungal nystatin and oral glutathione. Then I added binders like charcoal, bentonite (a type of clay), pectin, and cholestyramine, all of which can help keep mold neurotoxins in the gut until they can be eliminated. This was in addition to the IV glutathione and phosphatidylcholine mentioned above.

Katherine completed her course of antibiotics in March. By the first week in June, as the school year was coming to an end, Barbara was relieved to see a noticeable improvement in her daughter's demeanor and energy levels. Her color, eyes, and mood were brighter. Once again, Katherine went away to camp, and when she returned, there was no trace left of the sad, sickly girl she'd been before.

Mitochondrial Dysfunction

As we've seen, when the body's metabolic production gets compromised, it can lead to nutritional deficiencies or an inability for the body to detoxify, that is, remove and dispose of cellular waste and damage. These nutritional deficiencies and toxicants can lead to another type of metabolic imbalance in which the function of mitochondria becomes impaired, leading to significant loss of energy and cellular dysfunction.

As you might recall from high school biology, the mitochondria are our cellular "powerhouses," or the primary source for most of the energy—called adenosine triphosphate (ATP)—that fuels and regulates metabolism. In fact, the mitochondria produce 95 percent of the energy cells need to function. Primary mitochondrial dysfunction is caused by defects in the genes responsible for allowing the mitochondria to produce ATP. Secondary mitochondrial dysfunction also inhibits the mitochondria's ability to produce ATP, but it can be due to all kinds of other stressors that aren't directly related to the mitochondria. For example, the standard American diet could leave almost anyone nutritionally deficient or insufficient, a big problem given that the mechanisms that make ATP require adequate amounts of B vitamins, vitamin E, and Coenzyme Q10, to name just a few. Other stressors could include prenatal infections or complications, or childhood shocks and traumas. Certain seizure medicines, acetaminophen, and antibiotics could count as stressors, as could dehydration, fasting or starvation, surgery, bacterial and viral infections, and even stress itself. And yes, certain chemicals, heavy metals, and other environmental toxins can clog up the cellular machinery needed to make energy.

All organs need ATP to function, but the brain's energy requirements are the most of any, making it particularly vulnerable if its energy resources are disrupted. Several studies have suggested that low ATP production leads to neuronal death and neurodegeneration. In addition, evidence is building that the mitochondria plays a role in "neuronal development, growth, and regeneration." There is also a plethora of evidence to suggest that a combination of genetic and environmental factors can impact the mitochondria's contribution to "cellular resilience" and its ability to defend the brain against oxidative stress and prevent damage. So it's not surprising to learn that mitochondrial dysfunction can affect cognition and our moods. In fact, dysfunctions in this process are thought to play a role in the development of mental illnesses like bipolar disorder, depression, and schizophrenia.

Mitochondria are also concentrated in the blood-brain barrier, providing the energy needed to helps cells maintain their integrity. Mitochondrial dysfunction has been implicated in increased blood-brain barrier permeability, and researchers have linked mitochondrial disorders to mental health disorders like cognitive impairment, psychosis, and anxiety.

Mitochondrial dysfunction is an example of the systemic issues I look for when evaluating patients suffering from anxiety or depression (both bipolar, which manifests with extreme highs and lows, and unipolar, which manifests with low moods) who also exhibit some matching physical symptoms, such as low muscle tone, gastrointestinal symptoms, and extreme fatigue, especially during common illnesses. It's the kind of disorder that can be the tipping point for kids whose immune kettles are already bubbling close to the top.

To treat mitochondrial dysfunction, as always, improving one's diet will boost nutrient levels and help clear away toxins so the mitochondria can work better. I like to prescribe what I call a mitochondrial cocktail of high-level nutrients that may include vitamins B-2 and B-3, especially nicotinamide riboside, CoQ10 or ubiquinol, L-carnitine or acetyl-L-carnitine, magnesium, arginine, creatine, PQQ, vitamin C, vitamin E, and alpha-lipoic acid.

Adapt, Adjust, Repeat

Several of the case studies presented in this chapter followed patients as they grew from child to tween or tween to teen, revealing how the transition into adolescence can alter a child's equilibrium and force a reconsideration of treatment programs. A protocol that was once effective may need to be revisited over time.

As always, it's also important to remember that sometimes kids who've been on special diets and supplements for years forget how crummy they used to feel. Often they just want to be like the other kids. Or they're bored. Who the heck knows. The end result is that sometimes teenagers become less compliant and regress, and unfortunately it's impossible to know what pain point will be enough to get them back on track.

Emily doesn't have to take any meds at school, so there's no reason for her friends and classmates to know anything about her condition. Yet she often refuses to take her supplements. At one point she started to exhibit symptoms again, and Madeline found bags of supplements hidden all over the house. Emily finally admitted that she eats gluten when she's not around her parents, too. Rather than force the girl to take her pills in front of them like a prisoner, Madeline and Travis have decided to give her the freedom to decide for herself how to manage her health. They want her to learn any hard lessons when the consequences aren't too dire, since she's still living under their roof, where they have some hope of guiding her. Once she's living on her own they'll have no influence, and the last thing they want is for her to rebel when she goes away to college. There's no way to know, of course, if she's someone who will have to hit rock bottom before she decides to take control. Will she manage her health responsibly, or will she party hard and wreck her gut?

When you think about it, while the details may differ, Madeline and Travis are facing the exact same worry as any parent about a child preparing to leave the nest. Did we say the right things? Did we give her all the tools we possibly could have? As any parent of an adult child will tell you, just because your baby is all grown up and living on their own doesn't mean you ever stop worrying. The parent of any child who was once sick is going to be extra worried. Diane thinks maybe the early years of deprivation and overprotection may have led Jennifer to push back hard once she got away from home and her parents' eagle eyes. However, her partying in college eventually led her to realize she was using alcohol in unhealthy ways. She started seeing a therapist and got herself back on track. Today, at age twenty-seven, she chuckles when talking about her diet. Like any young professional, convenience is a cornerstone of her eating habits, and planning carefully prepared meals is not easy to do when you're working long days and still want time to have fun and be spontaneous. But she remembers what it was like to be sick, and she knows she feels better the more she sticks to a gluten-free, dairy-free, sugar-free lifestyle.

Eric is fourteen years old now. Postpuberty, he shot up another four inches and started filling out exactly as one would expect of a young teenage boy. His mother says he's extremely responsible about taking his supplements, but of course he still has a few years before reaching adulthood. Maybe he'll rebel. It's pretty rare to meet an eighteen- or nineteen-year-old who doesn't decide to have a beer every now and then. There's no way to know what kind of effect that could have on a sensitive boy like Eric. His parents talk to him about these things openly, however, believing it's better to help him work through his thoughts and answer his questions while they still might have some kind of influence. For now, though, there is reason for optimism. Recently, the boy who once couldn't handle any uncertainty willingly left home to spend a month sailing on a boat around the Caribbean.

COULD YOUR CHILD HAVE A NUTRITIONAL DEFICIENCY, METABOLIC DISORDER, OR SUFFER FROM TOXICITY?

Is your child's skin especially dry, and have you noticed any bumps on the outer part of your child's arms, or on the front of their thighs or cheeks?

Are there white spots on their nails?

Do you ever notice sores at the corner of your child's mouth?

Is your child eating a lot of tuna fish?

Do you live in a very old home, with old paint, and have you recently done a renovation?

Has there been any significant leaks in your home? Is there any visual mold? Is there a musty smell present?

Does your child get flushed easily, with red cheeks and/or ears? Is there a recurrence of strange rashes?

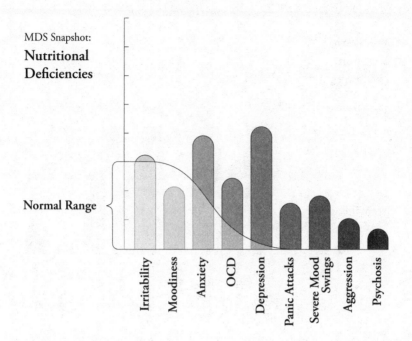

MDS Snapshot:
Nutritional Deficiencies

Normal Range

Irritability · Moodiness · Anxiety · OCD · Depression · Panic Attacks · Severe Mood Swings · Aggression · Psychosis

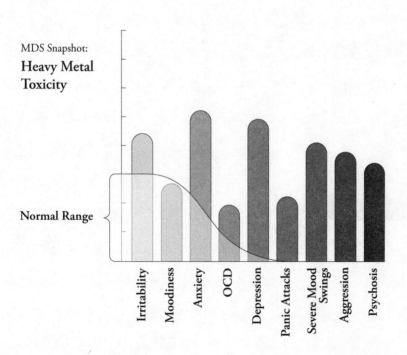

MDS Snapshot:
Heavy Metal Toxicity

Normal Range

Irritability · Moodiness · Anxiety · OCD · Depression · Panic Attacks · Severe Mood Swings · Aggression · Psychosis

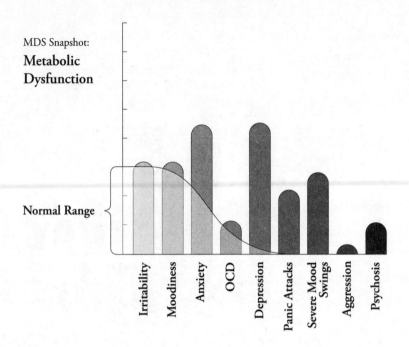

Conclusion

The first year after my residency, a patient I'd just seen spotted me eating lunch in a restaurant not far from the office. She ambled over to see what the "nutrition doctor" had on his plate. Her eyes widened when she saw. "Shrimp? They're so full of toxins!" she said. As politely as I could, I replied, "The amount of stress you're carrying over this shrimp is worse than anything this shrimp could do to you."

Maybe that attitude seems a little incongruous coming from a doctor who just spent so many pages chronicling possible complications from run-of-the-mill childhood illnesses like strep, listed so many possible allergens you're now wondering if there's anything that's safe to eat, and raised the specter that mold and other environmental toxins lurk in every nook and cranny of your house.

Remember, though, that every one of us is a product of a unique combination of genetic, environmental, neurologic, metabolic, hormonal, nutritional, and infectious influences. What affects one person dramatically won't bother the next at all. The world doesn't have to be a threatening place, especially when you're aware of the contents of your or your child's immune kettle and can prepare accordingly.

I didn't write this book to scare you, but to help you find answers and raise your awareness about factors you might not have considered before. And there is still so much to learn.

In medicine, as in other fields, we know what we know, we know what we don't know, and then there are all the things we don't know that we don't know (you got that?). For years I had an acupuncturist in my office because I could see that acupuncture eased my patients' pain, and sometimes even those suffering chronic immune-related illnesses. I had no idea how, but there was no doubt the practice helped. Acupuncturists swore it affected the immune system, but they couldn't say how or why. It had been used for thousands of years in Eastern medicine, but without contemporary studies to prove its efficacy, most Western doctors were reluctant to incorporate it into their treatments. Then scientists discovered that acupuncture modulates endorphins, which are one kind of messenger molecules for the nervous system. And T cells have receptors for endorphins. Aha! Now it made sense. If acupuncture modulates endorphins and T cells have receptors for endorphins, then acupuncture can affect T cells, which has a huge effect on the immune system.

With increasing scientific information as support, more and more Western doctors have prescribed acupuncture or endorsed their patients' interest in trying it. We know exponentially more about the brain, microbiome, immune system, and about how inflammation and external influences affect these systems than we did even five years ago. Every year we see more evidence of how different scientific disciplines like psychiatry, immunology, neurology, endocrinology, and gastroenterology are deeply interconnected. When we keep exploring and researching, the answers eventually reveal themselves.

It's gratifying to see how many of the treatments and ideas introduced by integrative physicians from when I first began my career are now being incorporated into mainstream medicine—such as the effect of the gut microbiome on our health and the existence of the gut-brain-immune axis. I expect the same thing will occur with many of the ideas presented in this book; this kind of progression is generally more the rule than the exception.

Information is empowering. Of course it can be confusing to see how much overlap there can be among the symptoms of brain inflammation, Lyme disease, allergies, thyroid and adrenal dysfunction, toxicity, mitochondrial dysfunction, and all the other disorders we discussed, when there's no typical presentation we can point to and say, "That's it!"

However, maybe you'll start to see a pattern. Maybe your child's MDS signature will bear a strong resemblance to one or two of the general models of ITABI, tick-borne disease, thyroid and/or adrenal dysfunction, allergies, sensitivities, nutritional deficiencies, or toxicity we've discussed, though any presentation will be unique and may also feature overlapping symptoms. Maybe you'll notice a detail in one of my patients' stories that you didn't realize was an important clue.

And maybe you won't. It's entirely possible, even likely, that after reading this book you'll be more confident that what's troubling your child isn't a medical issue at all, but purely psychosocial. In this case, they may very well ultimately find solace and healing with the help of a trusted therapist, or introduction to relaxation therapies, good sleep habits, and guidance in maneuvering the ups and downs of their online and offline social life and relationships. Either way, my hope is that you'll feel better knowing that you're aware of the options and possibilities, and were empowered to make the best educated choices on behalf of your child.

What's most important is that we don't let your child fall through the cracks. There are doctors who disagree with some of my approaches. Some of them will tell you the conditions I treat don't actually exist, or will refuse to consider your child could be suffering from them without a definitive lab result. But I've seen the positive impact of doing the detective work, of digging for the clues and comparing the symptoms, signs, history, triggers, and time lines to form a complete picture. Sometimes I hit a dead end. More often than not, though, I make a breakthrough. Sometimes that breakthrough has medical underpinnings; sometimes the breakthrough is that the issue is clearly psychosocial and beyond my expertise. Both scenarios are a win, as either conclusion puts my patient and their family on a new path, one that hopefully gets them a little closer to health and peace of mind.

Sometimes the journey to that new path can be long and arduous. It will be imperative that you take care of yourself physically and mentally along the way. You can't help your kids if you're running yourself into the ground with worry, looking for threats around every corner, sleeping poorly, or neglecting the relationships that sustain you. Loving, supportive, present parents are great preventive health care. We've talked a lot about supporting our children's microbiomes, immune systems, and cognitive and emotional development, but it's not just about what we feed them. The support of family, friends, and community is equally important.

We have to be flexible, making changes where we can in our lifestyles and in the products we choose to buy, while simultaneously forgiving ourselves when there are limits to what we can do. None of us can live in a sterile box, and of course that wouldn't be advisable anyway (or much fun). The best we can do is try to set or reset our children's immune kettles as low as they can go, then teach them to take the best care they can of their minds and bodies, so they can love, laugh, and live life fully, experiencing and enjoying the world as it is, not as we wish it would be. Remember, laughter and love are the two most powerful nourishments for your and your child's immune systems.

If you decide to pursue some of the approaches mentioned in these pages with the support of your trusted health-care provider, don't worry about doing everything perfectly. In an ideal world, we could all follow the advice in this book down to the last detail. Most people won't be able to, and that's okay, because this isn't an ideal world, it's the real one.

It's natural to feel overwhelmed when your child is suffering, and even when your child is healing. Some of those feelings are difficult. As parents start to see the return of the spark in their children's eyes and hear them laugh again, it's easy to mourn what's been lost. Time. Innocence. Precious traditions. A sense of safety and security. Taking the time to grieve is important. Allow yourself to acknowledge that there were times when you felt helpless, hopeless, when you lost patience or made mistakes. And then forgive yourself. I suggest you treat yourself with the same compassion that I would offer you if you were sitting in my office with your child. I would tell you that there's one thing I know for certain: no matter what you believe, you didn't screw up. Whatever is hurting your child is not your fault. Be gentle with yourself. If you're struggling with these feelings, you might try a simple exercise to help you access self-compassion: Place your hand on your heart, take a deep breath, and channel some of that immense, radiant love you feel for your child toward yourself.

This book is filled with stories of children and parents who've been where you are. I hope that hearing about their experiences will give you renewed energy and motivation to keep pushing to find the help your child needs. And remember: keep the communication lines between yourself and your child as open as possible no matter what. Above all, may these stories be a constant reminder that whatever it is you or your child are experiencing, you're not alone, and there is every reason to hope.

Acknowledgments

The many children, tweens, and teens I have seen over the course of the last few decades were the inspiration for this book. It is their struggles that moved me to find a literary agent who recognized the value of sharing their stories. Thank you, Laura Nolan, of Aevitas Creative Management, for your insight and guidance throughout this project. You have been a godsend. I especially thank you for helping me find Stephanie Land, an amazing writer and editor, who was invaluable in translating and synthesizing my knowledge, expertise, and clinical experience into accessible prose for the lay person, while also conducting interviews and masterfully weaving my patients' moving stories throughout the book. I know they will help change the trajectories of the lives of many children and adolescents. You were a pleasure to work with, and I appreciate all your efforts. Thank you, thank you, thank you, Stephanie.

I want to acknowledge my editor at HarperCollins Publishers, Julie Will, for her belief in and commitment to the message and importance of this book. Your editorial input, along with those of your assistant, Haley Swanson, were spot on and much appreciated.

I am grateful to one of my earliest mentors, George Engel, MD, who was a prominent psychiatrist at the University of Rochester School of Medicine and Dentistry. His teachings of compassion began during my earliest days of the first year. He was the father of biomedical psychosocial medicine and started the psychiatric-medicine liaison program at our medical school. His deep understanding of the bidirectional relationship between the mind and the body, coupled with his expertise in listening and his ability to nourish that in us, was something for which I will always be grateful. I believe his early teachings about the significance of the "open-ended interview" and the importance of paying attention to details in the physical exam set the foundation for my interest in pursuing answers to complex problems and propelled me to become a "medical detective."

Many thanks to Mathew den Boer, an incredible illustrator, who helped me create the visual representation of the Mood Dysregulation Spectrum (MDS), as well as expertly producing the other illustrations in the book. I also want to thank one of my closest friends, Kenny Solomon, as well as Michael Goode, for their initial help on one of the illustrations. Additionally, I am grateful to Kali Rosenblum for her help in making the self-compassion exercise so accessible.

I also want to thank my superb and dedicated staff, many of whom have been with me for twenty-to-thirty-plus years. They help me implement the strategies discussed in this book, and have played a major part in allowing me to help the many affected children, teens, tweens, and young adults that I have cared for over the course of my career. Their compassion, attention to detail, and follow-through have been a godsend for these patients. Thank you Theresa Scraggs; Barbara Coddington; Pam McCloskey, LPN; Michele Fletcher; Debbie Truin, RN; Valerie Marchini, RN; Jan Shufelt; Pam Wyant; Megan White; Tara Casey; and Veronica Mercuri, RN.

Last, I could never have written this book without the help and support of my family. Adding the researching and writing of a book onto a busy medical practice and lecture schedule put a lot more demands on my time, and I appreciate the understanding, love, and encouragement always offered by my wife, Marian, and my children Alicia and Jordan. You are the light of my life.

Appendix*

In the book I discuss numerous medications, using their generic or brand names at various times. I also discuss many nutrients and herbs, and mention the use of different combination formulas. These nutrients and herbs, both singularly and in various combinations, can be obtained from numerous sources. However, I thought it would be helpful to include here the combination formulas used in the case histories in the book. I provide brief descriptions below, but if you want more information about nutrients and nutritional approaches, you can visit bockintegrative.com or bocknutritionals.com.

SUGAR COMPANION

Key Ingredients:

Chromium, biotin, zinc

Promotes the healthy metabolism of carbohydrates
Designed to reduce sugar cravings

ALLERGY SUPPORT FORMULA

Key Ingredients:

Quercetin, vitamin C, bioflavonoids, stinging nettle extract

Supports individuals with immune imbalances and
    hypersensitivity reactions
Supports nasal and sinus passages

---

* These statements have not been evaluated by the Food and Drug Administration. These products are not intended to diagnose, treat, cure, or prevent any disease.

LIVER SUPPORT FORMULA

Key Ingredients:

Milk thistle seed extract (silymarin), N-acetyl cysteine (NAC), alpha-lipoic acid

Supports healthy liver function
Supports glutathione production
Supports antioxidant activity and healthy immune function

AB COMPANION

Key Ingredients:

Lactobacillus **and** *Bifidobacterium* **species,** *Saccharomyces boulardii*

Helps maintain a healthy intestinal flora
Supports gastrointestinal-based immunity
Supports intestinal barrier function and integrity

COBIOTIC PLUS

Key Ingredients:

Lactobacillus **and** *Bifidobacterium* **species,** *Saccharomyces boulardii*, **arabinogalactan**

Promotes a balanced intestinal flora
Supports immune health
Supports intestinal barrier function and integrity

EPI-GEN ACTIVATOR

Key Ingredients:

Curcumin, green tea, pterostilbene

Supports antioxidant activity
Supports natural detoxification pathways
Supports cytokine balance

RESVERATROL PRO

Key Ingredients:

Resveratrol, quercetin, pterostilbene

Provides antioxidant support
Supports healthy cellular function
Supports cardiovascular and neurological health

IMMUNE-UP PLUS

Key Ingredients:

Beta-glucan, olive leaf extract, vitamin C

Supports healthy immune function
Provides antioxidant support

D-3 5000 PLUS

Key Ingredients:

Vitamin D-3, vitamin K-2

Supports modulation of immune function
Supports neurological and cardiovascular health
Supports bone strength and dental health

GLUTAMINE PLUS POWDER

Key Ingredients:

L-glutamine, deglycyrrhizinated licorice extract, larch arabinogalactan extract, aloe vera extract

Supports gastrointestinal health
Helps maintain gut mucosal integrity and barrier function

METHYL SUPPORT

Key Ingredients:

Pyridoxal 5'-phosphate, methyl B-12, methyltetrahydrofolate, trimethylglycine

Provides bioactive form of B vitamins
Supports neurological and cardiovascular health
Supports maintenance of homocysteine levels
Supports healthy methylation and detoxification

MB-12 PLUS QD

Key Ingredients:

methylcobalamin, methyltetrahydrofolate

Provides the biologically active forms, methyl B-12 and
methyltetrahydrofolate
Supports methylation
Supports healthy nervous system function

THYROID SUPPORT FORMULA

Key Ingredients:

Thyroid extract, L-tyrosine, iodine, selenium

Provides support for healthy thyroid function

T3 SUPPORT FORMULA

Key Ingredients:

Selenium, iodine, zinc, vitamin A, guggul extract

Supports biosynthesis of the active T3 hormone

CORTISOL QUIET PLUS

Key Ingredients:

Magnolia officinalis, *Phellodendron amurense*, **and ashwagandha extracts, L-theanine**

Helps the body cope with the effects of stress
Supports healthy cortisol levels
Supports relaxation and restful sleep
Helps promote mental clarity

ADRENAL RELAX FORMULA

Key Ingredients:

Rhodiola, *Cordyceps*, **and Panax ginseng extracts**

Provides adaptogenic herbs
Helps support adrenal gland and immune function
Supports healthy energy levels

AG SUPPORT

Key Ingredients:

Glycyrrhiza, Panax ginseng, *Eleutherococcus, Schisandra, Rhodiola,* **adrenal extract**

Supports healthy adrenal function
Promotes energy production and stamina
Supports the body's response to stress

Notes

INTRODUCTION

xiv a 2016 Child Mind Institute report: Child Mind Institute, "2016 Children's Mental Health Report," Child Mind Institute, https://childmind.org/report /2016-childrens-mental-health-report/.

xiv According to the National Institute of Mental Health: "Mental Illness," National Institute of Mental Health, last updated February 2019, https://www .nimh.nih.gov/health/statistics/mental-illness.shtml#part_155771.

xiv Mental sluggishness: Manoj Kumar Sahoo, Ajit Avasthi, and Parampreet Singh, "Negative Symptoms Presenting as Neuropsychiatric Manifestation of Vitamin B12 Deficiency," *Indian Journal of Psychiatry* 53, no. 4 (October– December 2011): 370–71, https://doi.org/10.4103/0019-5545.91914.

xiv Panic attacks: James L. Schaller, Glenn A. Burkland, and P. J. Langhoff, "Do Bartonella Infections Cause Agitation, Panic Disorder, and Treatment-Resistant Depression?" *Medscape General Medicine* 9, no. 3 (September 2007): 54, https://www.ncbi.nlm.nih.gov/pmc/articles/PMC2100128/.

xv Depression, lethargy: "Mental Health," Celiac Disease Foundation, https:// celiac.org/about-celiac-disease/related-conditions/mental-health/.

xvi increasing around the turn of the millennium: "Data & Statistics on Autism Spectrum Disorder," Centers for Disease Control and Prevention, last reviewed March 25, 2020, https://www.cdc.gov/ncbddd/autism/data.html.

CHAPTER 1: WELCOME TO THE SPECTRUM

5 By the time they turn: "New Data on Mental Illness among U.S. Youth," *Monitor on Psychology* 42, no. 2 (2011): 10, https://www.apa.org/monitor /2011/02/mental-illness.aspx.

5 Five times as many: "Study: Youth Stress Exceeds Depression Era," CBSNews .com, January 11, 2010, https://www.cbsnews.com/news/study-youth-stress -exceeds-depression-era/.

5 One in five children: Child Mind Institute, "2016 Children's Mental Health Report," Child Mind Institute, https://childmind.org/report/2016-childrens -mental-health-report/.

5 Anxiety: Benoit Denizet-Lewis, "Why Are More American Teenagers Than Ever Suffering from Severe Anxiety?" *New York Times Magazine*, October 11, 2017, https://www.nytimes.com/2017/10/11/magazine/why-are-more -american-teenagers-than-ever-suffering-from-severe-anxiety.html.

6 Preliminary research correlates: Ye Li et al., "Dietary Patterns and Depression Risk: A Meta-analysis," *Psychiatry Research* 253 (July 2017): 373–82, https:// doi.org/10.1016/j.psychres.2017.04.020.

6 One out of four: Bill Hendrick, "Most Young Kids Don't Get Enough Exer- cise," WebMD.com, April 4, 2011, https://www.webmd.com/children/news /20110414/most-young-kids-dont-get-enough-exercise#1.

6 and many teenagers don't get: Alice Park, "Teens Are Just as Sedentary as 60-Year-Olds," *Time*, June 16, 2017, https://time.com/4821963/teens -sedentary-lifestyle-exercise/.

6 Children under the age: Stacy Hodgkinson et al., "Improving Mental Health Access for Low-Income Children and Families in the Primary Care Setting," *Pediatrics* 139, no. 1 (January 2017): https://doi.org/10.1542/peds.2015 -1175.

6 Youth living in two: Reed Jordan, "Poverty's Toll on Mental Health," Urban Institute, November 25, 2013, https://www.urban.org/urban-wire/povertys -toll-mental-health.

6 Yet kids from higher-income environments: Suniya S. Luthar and Shawn J. Latendresse, "Children of the Affluent: Challenges to Well-Being," *Current Directions in Psychological Science* 14, no. 1 (2005): 49–53, https://doi.org /10.1111/j.0963-7214.2005.00333.x.

6 Distant relationships: Luthar and Latendresse, "Children of the Affluent," 49–53.

7 Uncompromising high-pressure: Suniya S. Luthar, "The Problem with Rich Kids," *Psychology Today*, November 2013, https://www.psychologytoday.com /us/articles/201311/the-problem-rich-kids.

7 In 2018, Gen Z teens: American Psychological Association, "Stress in America: Generation Z," *Stress in America Survey* (2018): 3, https://www.apa.org/news /press/releases/stress/2018/stress-gen-z.pdf.

7 Philip Kendall, director: Amy Ellis Nutt, "Why Kids and Teens May Face Far More Anxiety These Days," *Washington Post*, May 10, 2018, https:// www.washingtonpost.com/news/to-your-health/wp/2018/05/10/why-kids -and-teens-may-face-far-more-anxiety-these-days/?noredirect=on&utm_term =.29a77a8ac9f5.

7 more than 250,000 Americans: CDC, "Coronavirus Disease 2019 (Covid-19) Cases and Deaths in the U.S.," Centers for Disease Control and Prevention, https://www.cdc.gov/coronavirus/2019-ncov/cases-updates/us-cases-deaths .html.

7 With today's adolescents already reporting: Jean M. Twenge, Brian H. Spitzberg, and W. Keith Campbell, "Less In-Person Social Interaction with Peers among U.S. Adolescents in the 21st century and Links to Loneliness," *Journal of Social and Personal Relationships* 36, no. 6 (March 2019), https://doi .org/10.1177/0265407519836170.

7 teen mental health expert Jean Twenge: Juliana Kaplan, "Because of the Pandemic, Gen Z Will Likely See Its Mental Health Deteriorate Even Further—and Social Lives Will Get Even More Insular," *Business Insider*, May 12, 2020, https://www.businessinsider.com/gen-z-mental-health-will-deteriorate-social -lives-more-insular-2020-5.

7 the role social media: "Mental Health Issues Increased Significantly in Young Adults Over Last Decade," American Psychological Association, March 14, 2019, https://www.apa.org/news/press/releases/2019/03/mental-health -adults.

7 Between 2007 and 2012: Amy Ellis Nutt, "Why Kids and Teens May Face Far More Anxiety These Days," *Washington Post* , May 10, 2018, https://www .washingtonpost.com/news/to-your-health/wp/2018/05/10/why-kids-and -teens-may-face-far-more-anxiety-these-days/.

8 Fifty percent more teenagers: Jean M. Twenge, PhD, "Why So Many Teens Today Have Become Depressed," *Psychology Today*, August 25, 2017, https:// www.psychologytoday.com/us/blog/our-changing-culture/201708/why-so -many-teens-today-have-become-depressed.

8 Says Janis Whitlock: Susanna Schrobsdorff, "Teen Depression and Anxiety: Why the Kids Are Not Alright," *Time*, October 27, 2016, https://time.com /magazine/us/4547305/november-7th-2016-vol-188-no-19-u-s/.

13 witnessing threats to the ones we love: Frank Muscara et al., "Parent Distress Reactions Following a Serious Illness or Injury in Their Child: A Protocol Paper for the Take a Breath Cohort Study," *BMC Psychiatry* 153, no. 15 (2015), https://doi.org/10.1186/s12888-015-0519-5.

CHAPTER 2: IMMUNE SYSTEM 101

19 kynurenic acid can't be metabolized: Matthew A. Ciorba, "Kynurenine Pathway Metabolites: Relevant to Vitamin B-6 Deficiency and Beyond," *American Journal of Clinical Nutrition* 98, no. 4 (2013): 863–64, https://doi.org/10.3945/ajcn.113.072025.

19 vitamin B-6 insufficiency: Sara K. Olsson et. al., "Elevated Levels of Kynurenic Acid Change the Dopaminergic Response to Amphetamine: Implications for Schizophrenia," *International Journal of Neuropsychopharmacology* 12, no. 4 (2009): 501–12, https://doi.org/10.1017/S1461145708009383; Norbert Müller, Aye-Mu Myint, and Markus J. Schwarz, "Kynurenine Pathway in Schizophrenia: Pathophysiological and Therapeutic Aspects," *Current Pharmaceutical Design* 17, no. 2 (2011): 130–36, https://doi.org/10.2174/138161211795049552.

21 Until the early 1990s, U.S. environmental policy: Philip J. Landrigan and Lynn R. Goldman, "Children's Vulnerability to Toxic Chemicals: A Challenge and Opportunity to Strengthen Health and Environmental Policy," *Health Affairs* 30, no. 5 (2011), https://doi.org/10.1377/hlthaff.2011.0151.

21 The Environmental Protection Agency (EPA) lists: "TSCA Chemical Substance Inventory," United States Environmental Protection Agency, accessed July 2020, https://www.epa.gov/tsca-inventory/about-tsca-chemical-substance-inventory.

21 most have never been thoroughly safety-tested: Ian Urbina, "Think Those Chemicals Have Been Tested?," *New York Times*, April 13, 2013, https://www.nytimes.com/2013/04/14/sunday-review/think-those-chemicals-have-been-tested.html; Richard Denison, "5 Ways Chemical Safety Is Eroding Under Trump," Environmental Defense Fund, last updated May 13, 2019, https://www.edf.org/blog/2018/05/01/5-ways-chemical-safety-eroding-under-trump.

23 heart disease to cancer to diabetes: Meghana Keshavan, "Tantalizing Clues Point to Inflammation's Role in an Array of Diseases," *Scientific American*, September 1, 2017, https://www.scientificamerican.com/article/tantalizing-clues-point-to-inflammations-role-in-an-array-of-diseases/.

23 inflammation contributes to autoimmunity: Andrew H. Miller, Ebrahim Haroon, and Jennifer C. Felger, "Therapeutic Implications of Brain-Immune Interactions: Treatment in Translation," *Neuropsychopharmacology* 42, no.1 (2017): 334–59, https://doi.org/10.1038/npp.2016.167.

24 some anti-inflammatory pain relievers: Rickinder Sethi et al., "Neurobiology and Therapeutic Potential of Cyclooxygenase-2 (COX-2) Inhibitors for In-

flammation in Neuropsychiatric Disorders," *Frontiers in Psychiatry* 605, no. 10 (2019), https://doi.org/10.3389/fpsyt.2019.00605.

27 That lack of communication: Tori Rodriguez, "How the Immune System Influences Psychiatric Disorders," *Psychiatry Advisor*, November 3, 2015, https://www.psychiatryadvisor.com/home/topics/mood-disorders/how-the-immune-system-influences-psychiatric-disorders/.

27 The CNS was thought to be "immune privileged": Antoine Louveau, Tajie H. Harris, and Jonathan Kipnis, "Revisiting the Mechanisms of CNS Immune Privilege," *Trends in Immunology* 36, no. 10 (2015): 569–77, https://doi.org/10.1016/j.it.2015.08.006.

27 inflammatory cytokines: Tori Rodriguez, "How the Immune System Influences," *Psychiatry Advisor*, November 3, 2015, https://www.psychiatryadvisor.com/home/topics/mood-disorders/how-the-immune-system-influences-psychiatric-disorders/.

27 an association between certain inflammatory markers: Rita Haapakoski et al., "Cumulative Meta-Analysis of Interleukins 6 and 1β, Tumour Necrosis Factor α and C-reactive Protein in Patients with Major Depressive Disorder," *Brain, Behavior, and Immunity* 49 (2015): 206–15, https://doi.org/10.1016/j.bbi.2015.06.001.

29 Powerful inflammatory signaling: Narayanan Parameswaran and Sonika Patial, "Tumor Necrosis Factor-α Signaling in Macrophages," *Critical Reviews in Eukaryotic Gene Expression* 20, no. 2 (2010): 87–103. https://doi.org/10.1615/critreveukargeneexpr.v20.i2.10.

30 often in as little as thirty minutes: Sofia de Oliveira, Emily E. Rosowski, and Anna Huttenlocher, "Neutrophil Migration in Infection and Wound Repair: Going Forward in Reverse," *Nature Reviews Immunology* 16, no. 6 (2016): 378–91, https://doi.org/10.1038/nri.2016.49.

30 symptoms like hay fever: Kensuke Miyake and Hajime Karasuyama, "Emerging Roles of Basophils in Allergic Inflammation," *Allergology International* 66, no. 3 (2017): 382–91, https://doi.org/10.1016/j.alit.2017.04.007.

31 The dendritic cells essentially guide the T and B cells: Maria Rescigno and Antonio Di Sabatino, "Dendritic Cells in Intestinal Homeostasis and Disease," *Journal of Clinical Investigation* 1, no. 9 (2009): 2441–50, https://doi.org/10.1172/JCI39134.

31 They help maintain: Richard Watson, et al., "An Introduction to Immunology and Immunopathology," *Allergy, Asthma & Clinical Immunology* 7, no. 1 (2011): 51–58, https://doi.org/10.1186/1710-1492-7-S1-S1.

31 "physiologically stressed": Sourav Paul and Girdhari Lal, "The Molecular

Mechanism of Natural Killer Cells Function and Its Importance in Cancer Immunotherapy," *Frontiers in Immunology* 1124, no. 8 (2017), https://doi .org/10.3389/fimmu.2017.01124.

31 producing pro-inflammatory: Jonas Schluter and Kevin R. Foster, "The Evolution of Mutualism in Gut Microbiota via Host Epithelial Selection," *PLOS Biology* 10, no. 11 (2012): https://doi.org/10.1371/journal.pbio.1001424.

32 Although most plasma cells die: Holger W. Auner, et al., "The Life Span of Short-Lived Plasma Cells Is Partly Determined by a Block on Activation of Apoptotic Caspases Acting in Combination with Endoplasmic Reticulum Stress," *Blood* 116, no. 18 (2010): 3445–55, https://doi.org/10.1182/blood -2009-10-250423.

35 resulting inflammation: Matthew S. Kayser and Josep Dalmau, "The Emerging Link Between Autoimmune Disorders and Neuropsychiatric Disease," *Journal of Neuropsychiatry and Clinical Neurosciences* 23, no. 1 (2011): 90–97, https://doi.org/10.1176/appi.neuropsych.23.1.90; Frank Leypoldt, Thaís Armangue, and Josep Dalmau, "Autoimmune Encephalopathies," *Annals of the New York Academy of Sciences*, 1338, no. 1 (2015): 94–114, https://doi .org/10.1111/nyas.12553.

36 Our environment is increasingly: Huimin Yan, Masaya Takamoto, and Kazuo Sugane, "Exposure to Bisphenol A Prenatally or in Adulthood Promotes TH2 Cytokine Production Associated with Reduction of CD4+CD25+ Regulatory T Cells," *Environmental Health Perspectives* 116, no. 4 (2008): 514–19, https://ehp.niehs.nih.gov/doi/pdf/10.1289/ehp.10829.

36 evidence shows that: J. Rodrigo Mora, Makoto Iwata, and Ulrich H. von Andrian, "Vitamin Effects on the Immune System: Vitamins A and D Take Centre Stage," *Nature Reviews Immunology* 8, no. 9 (2008): 685–98, https:// doi.org/10.1038/nri2378.

36 and Vitamin D: Chen-Yen Yang et al., "The Implication of Vitamin D and Autoimmunity: A Comprehensive Review," *Clinical Reviews in Allergy & Immunology* 45, no. 2 (2013): 217–26, https://doi.org/10.1007/s12016-013 -8361-3.

36 infections: Michael E. Benros et al., "Autoimmune Diseases and Severe Infections as Risk Factors for Mood Disorders: A Nationwide Study," *JAMA Psychiatry* 70, no. 8 (2013): 812–20, https://doi.org/10.1001/jamapsychiatry .2013.1111; Rosalie Greenberg, "The Role of Infection and Immune Responsiveness in a Case of Treatment-Resistant Pediatric Bipolar Disorder," *Frontiers in Psychiatry* 8, no. 78 (2017): https://doi.org/10.3389/fpsyt.2017 .00078; Martin F. Bachmann and Manfred Kopf, "On the Role of the Innate

Immunity in Autoimmune Disease," *Journal of Experimental Medicine* 193, no. 12 (2001): 147–50, https://doi.org/10.1084/jem.193.12.f47; Crystal Phend, "Depression Tied to Infections, Autoimmune Disease," MedPage Today, last updated June 13, 2013, https://www.medpagetoday.com/psychiatry/depression /39794.

36 In utero: Adina Ziemba-Goldfarb, "Possible Mechanisms That Protect the Fetus from Maternal Rejection," *Science Journal of the Lander College of Arts and Sciences* 10, no. 2 (Spring 2017): 56–61, http://touroscholar.touro.edu /cgi/viewcontent.cgi?article=1008&context=sjlcas.

36 My colleague Sid Baker: Sid Baker, interview with Dave Asprey, Bullet-proof Radio, May 2016, https://blog.daveasprey.com/wp-content/uploads /2016/05/Transcript-Sid-Baker.pdf.

CHAPTER 3: GUT FEELINGS

42 such as religious fervor: Thiago Cardoso Vale and Francisco Cardoso, "Chorea: A Journey Through History," *Tremor and Other Hyperkinetic Movements* 5 (2015):1–6, https://doi.org/10.7916/D8WM1C98.

42 bad weather and poor harvests: John Waller, "A Forgotten Plague: Making Sense of Dancing Mania," *The Lancet* 373, no. 9664 (February 2009): 624–25, https://doi.org/10.1016/S0140-6736(09)60386-X.

42 mass "dancing mania:" Vale and Cardoso, "Chorea."

42 and even death: Patricia Bauer, "Dancing Plague of 1518," Britannica.com, May 18, 2017, https://www.britannica.com/event/dancing-plague-of-1518.

42 Some theorize that toxic levels of ergot: Linnda R. Caporael, "Ergotism: The Satan Loosed in Salem?," *Science* 192 (April 1976), http://www.physics. smu.edu/scalise/P3333sp08/Ulcers/ergotism.html; Torbjørn Alm and Brita Elvevåg, "Ergotism in Norway. Part 1: The Symptoms and Their Interpretation from the Late Iron Age to the Seventeenth Century," *History of Psychiatry* 24, no. 1 (2012): 19, https://doi.org/10.1177/0957154X11433960.

42 Religious pressure and superstition: Stacy Schiff, *The Witches* (Boston: Little, Brown and Co., 2015), 22, 130–39, 386–93, 397–99.

42 ran a lab: "Ilya Mechnikov," *Nobel Lectures, Physiology or Medicine 1901–1921* (Amsterdam: Elsevier Publishing Co., 1967), https://www.nobelprize.org /prizes/medicine/1908/mechnikov/biographical/.

42 sparked a run on yogurt: Luba Vikhanski, "A Science Lecture Accidentally Sparked a Global Craze for Yogurt," *Smithsonian Magazine*, April 11, 2016, https:// www.smithsonianmag.com/science-nature/science-lecture-accidentally -sparked-global-craze-yogurt-180958700/.

42 Four years later: "Ilya Mechnikov," *Nobel Lectures, Physiology or Medicine 1901–1921*.

43 it was once thought: Alison Abbott, "Scientists Bust Myth That Our Bodies Have More Bacteria Than Human Cells," *Nature*, January 2016, https://doi .org/10.1038/nature.2016.19136.

43 small molecules produced by: Maayan Levy, Christoph A. Thaiss, and Eran Elinav, "Metabolites: Messengers Between the Microbiota and the Immune System," *Genes and Development* 30 (2016): 1589–97, https://doi.org /10.1101/gad.284091.116.

44 about four pounds' worth: Paul Forsyth and Wolfgang A. Kunze, "Voices from Within: Gut Microbes and the CNS?," *Cellular and Molecular Life Sciences*, 70, no. 1 (2013): 55–69, https://doi.org/10.1007/s00018-012-1028-z.

44 with the majority residing: R. Sender, and R. Milo, S. Fuchs, "Revised Estimates for the Number of Human and Bacteria Cells in the Body," *PLOS Biology*, August 19, 2016, https://doi.org/10.1371/journal.pbio.1002533.

44 the microbiome is so vast: Michael C. Toh and Emma Allen-Verco, "The Human Gut Microbiota with Reference to Autism Spectrum Disorder: Considering the Whole as More Than a Sum of Its Parts," *Microbial Ecology in Health and Disease* 26 (2015), https://www.tandfonline.com/doi/full/10.3402/mehd .v26.26309.

45 Tennis court: Jesse D. Aitken and Andrew T. Gewirtz, "Toward Understanding and Manipulating the Gut Microbiota," *Nature Reviews Gastroenterology & Hepatology* 10 (2013): 7274, https://doi.org/10.1038/nrgastro.2012.252.

47 500 million neurons: Forsyth and Kunze, "Voices from Within."

47 "second brain:" Michael D. Gershon, "The Enteric Nervous System: A Second Brain," *Hospital Practice* 34, no. 7 (1999): 31–52, https://doi.org/10.3810 /hp.1999.07.153.

47 "emotional and cognitive": Marilia Carabotti et al., "The Gut-Brain Axis: Interactions Between Enteric Microbiota, Central and Enteric Nervous Systems," *Annals of Gastroenterology* 28, no. 2 (2015): 203–9, https://www.ncbi .nlm.nih.gov/pmc/articles/PMC4367209/.

47 vagus nerve: Sigrid Breit et al., "Vagus Nerve as Modulator of the Brain-Gut Axis in Psychiatric and Inflammatory Disorders," *Frontiers in Psychiatry* 9 no. 44 (2018), 10.3389/fpsyt.2018.00044.

48 alter the production of: Carabotti et al., "The Gut-Brain Axis."

48 a "community" of bacteria: K. A. Floyd et al., "Adhesion of Bacteria to Surfaces and Biofilm Formation on Medical Devices," in *Biofilms and Implantable Medical Devices*, eds. Ying Deng and Wei Lv (Sawston, UK: Woodhead Pub-

lishing, 2017), 47–95, https://www.sciencedirect.com/topics/engineering/biofilm-community.

48 The microbiome controls: Daniel Erny et al., "Host Microbiota Constantly Control Maturation and Function of Microglia in the CNS," *Nature Neuroscience* 18 no. 7 (2015): 965–77, https://doi.org/10.1038/nn.4030.

48 Gut microbes also produce neurotransmitters: Carabotti et al., "The Gut-Brain Axis."

48 About *90 percent*: Jessica Stoller-Conrad, "Microbes Help Produce Serotonin in Gut," CalTech.edu, April 9, 2015, https://www.caltech.edu/about/news/microbes-help-produce-serotonin-gut-46495.

48 the BBB provides: K. K. Jain, "Blood-Brain Barrier," *Medlink Neurology*, updated July 2019, https://www.medlink.com/article/blood-brain_barrier.

49 plenty of disrupters: Jay P. Patel and Benicio N. Frey, "Disruption in the Blood-Brain Barrier: The Missing Link Between Brain and Body Inflammation in Bipolar Disorder?," *Neural Plasticity*, Article ID 708306 (2015), https://doi.org/10.1155/2015/708306; Birgit Obermeier, Richard Daneman, and Richard M. Ransohoff, "Development, Maintenance and Disruption of the Blood-Brain Barrier," *Nature Medicine* 19, no. 12 (2013): 1584–96, https://doi.org/10.1038/nm.3407.

49 radiation: Robert A. Nordal and C. Shun Wong, "Molecular Targets in Radiation-Induced Blood-Brain Barrier Disruption," *International Journal of Radiation Oncology* 62, no. 1 (2005): 279–87, https://doi.org/10.1016/j.ijrobp.2005.01.039.

50 brain injury: Himakarnika Alluri et al., "Blood-Brain Barrier Dysfunction Following Traumatic Brain Injury," *Metabolic Brain Disease* 30, no. 5 (2015): 1093–1104, https://doi.org/10.1007/s11011-015-9651-7.

50 low in protein: C. C. de Aquino et al., "Effect of Hypoproteic and High-Fat Diets on Hippocampal Blood-Brain Barrier Permeability and Oxidative Stress," *Frontiers in Nutrition* 5, no. 131 (January 2019), https://doi.org/10.3389/fnut.2018.00131.

50 glymphatic system: "Glymphatic System," University of Rochester Medical Center, Nedergaard Lab, Lab Focuses, https://www.urmc.rochester.edu/labs/nedergaard/projects/glymphatic-system.aspx; Natalie L. Hauglund, Chiara Pavan, and Maiken Nedergaard, "Cleaning the Sleeping Brain: The Potential Restorative Function of the Glymphatic System," *Current Opinion in Physiology* 15 (June 2020): 1–6, https://doi.org/10.1016/j.cophys.2019.10.020.

50 The American Academy of Sleep Medicine: Shalini Paruthi et al., "Consensus Statement of the American Academy of Sleep Medicine on the Recommended

Amount of Sleep for Healthy Children: Methodology and Discussion," *Journal of Clinical Sleep Medicine* 12, no. 11 (November 2016): 1549–61, https://doi.org/10.5664/jcsm.6288.

50 researchers claim teenagers need: "Sleep in Adolescents," Nationwide Children's Hospital, Sleep Disorder Center, https://www.nationwidechildrens.org/specialties/sleep-disorder-center/sleep-in-adolescents.

50 What we learned is that large particles can make it through the BBB: Antoine Louveau, Tajie H. Harris, and Jonathan Kipnis, "Revisiting the Mechanics of CNS Immune Privilege," *Trends in Immunology* 36, no. 10 (2015): https://doi.org/10.1016/j.it.2015.08.006.

51 In the lab: M. L. Wong, et al., "Inflammasome Signaling Affects Anxiety- and Depressive-Like Behavior and Gut Microbiome Composition," *Molecular Psychiatry* 2, no. 6 (2016): 797–805, https://doi.org/10.1038/mp.2016.46.

51 Associations between: Emeran A. Mayer et al. "Gut/Brain Axis and the Microbiota," *Journal of Clinical Investigation* 125 no. 3 (2015), 926–38, https://doi.org/10.1172/JCI76304; Francesca Mangiola et al., "Gut Microbiota in Autism and Mood Disorders," *World Journal of Gastroenterology* 22, no. 1 (2016): 361–68, https://doi.org/10.3748/wjg.v22.i1.361.

51 Some of the more common psychological side effects: G. B. Rogers et al., "From Gut Dysbiosis to Altered Brain Function and Mental Illness: Mechanisms and Pathways," *Molecular Psychiatry* 21 (April 2016): 738–48, https://doi.org/10.1038/mp.2016.50.

51 When we treat the fungal dysbiosis: Ger Bongaerts, René Severijnen, and Harro Timmerman, "Effect of Antibiotics, Prebiotics and Probiotics in Treatment for Hepatic Encephalopathy," *Medical Hypotheses* 64, no. 1 (2005): 64–68, https://doi.org/10.1016/j.mehy.2004.07.029.

CHAPTER 4: OUT OF CONTROL

62 Some researchers even suspect: "Pediatric Autoimmune Neuropsychiatric Disorders Associated with Streptococcus Infections," National Institutes of Health, last updated July 1, 2020, https://rarediseases.info.nih.gov/diseases/7312/pediatric-autoimmune-neuropsychiatric-disorders-associated-with-streptococcus-infections.

62 Autoimmune diseases are among the most common: NIH Autoimmune Diseases Coordinating Committee, "Progress in Autoimmune Diseases Research," National Institutes of Health (March 2005), accessed July 19, 2019, https://www.niaid.nih.gov/sites/default/files/adccfinal.pdf.

62 the American Autoimmune Related: "Autoimmune Disease . . . ," American

Autoimmune Related Diseases Association, accessed July 2019, https://www
.aarda.org/news-information/statistics/.

64 The study grew from work: "Dr. Judith Rapoport Oral History 2017," interview
by Ramya Rajagopalan, National Institutes of Health, August 28, 2017, https://
history.nih.gov/display/history/Rapoport%2C+Judith+2017; Susan Swedo,
"Interview with Dr. Susan Swedo," interview by Anna Conkey, Neuroimmune
.org, Foundation for Children with Neuroimmune Disorders, March 2018,
https://www.neuroimmune.org/susan-swedo-pandas-interview/.

64 A few months later: Susan E. Swedo, James F. Leckman, and Noel R. Rose,
"From Research Subgroup to Clinical Syndrome: Modifying the PANDAS
Criteria to Describe PANS (Pediatric Acute-Onset Neuropsychiatric Syndrome)," *Pediatrics and Therapeutics* 2, no. 2 (2012): 113, https://doi.org
/10.4172/2161-0665.1000113.

65 In 1998: Susan S. E. Swedo et al., "Pediatric Autoimmune Neuropsychiatric
Disorders Associated with Streptococcal Infections: Clinical Description of
the First 50 Cases," *American Journal of Psychiatry* 155, no. 2 (February 1998):
264–71.

67 In 2012: Swedo, "From Research Subgroup," 113.

68 pediatricians: Jennifer Frankovich et.al., "Clinical Management of Pediatric
Acute-Onset Neuropsychiatric Syndrome," *Journal of Child and Adolescent
Psychopharmacology* 27, no. 7 (September 2017): 574–93, http://doi.org/10.1089
/cap.2016.0148.

69 "environmental mosaic": Shaye Kivity et al., "Infections and Autoimmunity—
Friends or Foes?," *Trends in Immunology* (July 2009), https://doi.org/10.1016/j
.it.2009.05.005.

69 mycoplasma pneumonia: N. Muller et al., "*Mycoplasma Pneumoniae* Infection and Tourette's Syndrome," *Psychiatry Research* 129, no. 2 (December
2004): 119–25, https://doi.org/10.1016/j.psychres.2004.04.009.

69 chicken pox: R. C. Dale, A. J. Church, and I. Heyman, "Striatal Encephalitis after Varicella Zoster Infection Complicated by Tourettism," *Movement
Disorders: Official Journal of the International Parkinson and Movement Disorder Society* 18, no. 12 (December 2003): 1554–56, https://doi.org/10.1002
/mds.10610.

69 common cold: P. J. Hoekstra et al., "Association of Common Cold with Ex-
acerbations in Pediatric but Not Adult Patients with Tic Disorder: A Prospec-
tive Longitudinal Study," *Journal of Child and Adolescent Psychopharmacology*
15, no. 2 (April 2005): 285–92, https://doi.org/10.1089/cap.2005.15.285.

75 first reported by dermatologists: I. K. Aronson and K. Soltani, "The Enigma

of the Pathogenesis of the Jarisch-Herxheimer Reaction," *British Journal of Venereal Diseases* 52, no. 5 (October 1976): 313–15, https://doi.org/10.1136 /sti.52.5.313.

77 working with autistic children: Paul Ashwood et al., "Spontaneous Mucosal Lymphocyte Cytokine Profiles in Children with Autism and Gastrointestinal Symptoms: Mucosal Immune Activation and Reduced Counter Regulatory Interleukin-10," *Journal of Clinical Immunology* 24 (November 2004), 664–73, https://doi.org/10.1007/s10875-004-6241-6.

77 a strict ketogenic diet: H. Guo, J. Callaway, and J. Ting, "Inflammasomes: Mechanism of Action, Role in Disease, and Therapeutics," *Nature Medicine* 21 (2015): 677–87, https://doi.org/10.1038/nm.3893.

78 also showed promise: E. Brietzke et al., "Ketogenic Diet as a Metabolic Therapy for Mood Disorders: Evidence and Developments," *Neuroscience & Biobehavioral Reviews* 94 (2018): 11–16, doi:10.1016/j.neubiorev.2018.07.020.

79 The World Health Organization has stated: 39th Expert Committee on Drug Dependence, "Cannabidiol (CBD) Pre-Review Report," World Health Organization, Agenda Item 5.2 (November 2017): 5, https://www.who.int/medicines /access/controlled-substances/5.2_CBD.eps.

79 which help regulate cells: J. Komorowski and H. Stepień, "The Role of the Endocannabinoid System in the Regulation of Endocrine Function and in the Control of Energy Balance in Humans," *Postepy Higieny I Medycyny Doswiadczalnej* 61 (2007): 99–105, https://www.ncbi.nlm.nih.gov/pubmed /17369778.

79 and gastrointestinal tract: A. A. Izzo and A. A. Coutts, "Cannabinoids and the Digestive Tract," in *Cannabinoids: Handbook of Experimental Pharmacology*, ed. R. G. Pertwee (New York: Springer-Verlag, 2005), 573–98, https://doi .org/10.1007/3-540-26573-2_19.

79 including the enteric nervous system: Izzo and Coutts, "Cannabinoids," 573–98.

79 also mast cells: Maria-Teresa Samson et al., "Differential Roles of CB1 and CB2 Cannabinoid Receptors in Mast Cells," *Journal of Immunology* 170, no. 10 (May 2003): 4953–62, 10.4049/jimmunol.170.10.4953.

79 microglia in the brain: Debra A. Kendall and Guillermo A. Yudowski, "Cannabinoid Receptors in the Central Nervous System: Their Signaling and Roles in Disease," *Frontiers in Cellular Neuroscience* (January 2017), https://doi.org /10.3389/fncel.2016.00294.

79 found in the peripheral nervous system: Kendall and Yudowski, "Cannabinoid."

79 "runner's high": Johannes Fuss et al., "A Runner's High Depends on Canna-

binoid Receptors in Mice," *Proceedings of the National Academy of Sciences* 112, no. 42 (October 2015): 13105–108, https://doi.org/10.1073/pnas.1514996112.

80 mood boost: Emmanuelle di Tomaso, Massimiliano Beltramo, and Daniele Piomelli, "Brain Cannabinoids in Chocolate," *Nature* 382 (August 1996): 677–78, https://doi.org/10.1038/382677a0.

80 Anandamide levels: G. B. Stefano, et al., "Endocannabinoids as Autoregulatory Signaling Molecules: Coupling to Nitric Oxide and a Possible Association with the Relaxation Response," *Medical Science Monitor* 9, no. 4, (2003): RA63–75.

80 acupuncture: Lin Chen et al., "Endogenous Anandamide and Cannabinoid Receptor-2 Contribute to Electroacupuncture Analgesia in Rats," *Journal of Pain* 10, no. 7 (July 2009): 732–39, https://doi.org/10.1016/j.jpain.2008.12.012.

80 Stimulating synaptic plasticity: James M. Nichols and Barbara L.F. Kaplan, "Immune Responses Regulated by Cannabidiol," *Cannabis and Cannabinoid Research*, 5, no. 1 (March 2020): 12–31, https://doi.org/10.1089/can.2018.0073.

80 a synergism: Shimon Ben-Shabat et al., "An Entourage Effect: Inactive Endogenous Fatty Acid Glycerol Esters Enhance 2-Arachidonoyl-Glycerol Cannabinoid Activity," *European Journal of Pharmacology* 353, no. 1 (July 17, 1998): 23–31, https://doi.org/10.1016/s0014-2999(98)00392-6.

81 antibodies derived from: S. Jolles, W. A. Sewell, and S. A. Misbah, "Clinical Uses of Intravenous Immunoglobulin," *Clinical & Experimental Immunology* 142, no. 1 (2005): 1–11, doi:10.1111/j.1365-2249.2005.02834.x.

CHAPTER 5: BITTEN BY THE BUG

88 The story began: Boyce Rensberger, "A New Type of Arthritis Found in Lyme," *New York Times*, July 18, 1976, https://www.nytimes.com/1976/07/18/archives/a-new-type-of-arthritis-found-in-lyme-new-form-of-arthritis-is.html.

88 Polly was familiar: Jonathan A. Edlow, *Bullseye: Unraveling the Medical Mystery of Lyme Disease* (New Haven, CT.: Yale University Press, 2003): 1–2, https://books.google.com/books?hl=en&lr=&id=SokNhmzFERIC&oi=fnd&pg=PR9&ots=KsocJmc8R2&sig=mrTWhBNUazC7MKcuiXqCFijMvr0#v=onepage&q&f=false.

88 The symptoms did look: Rensberger, "A New Type of Arthritis."

88 In 1978: A. C. Steere, T. F. Broderick, and S. E. Malawista, "Erythema Chronicum Migrans and Lyme Arthritis: Epidemiologic Evidence for a Tick

Vector," *American Journal of Epidemiology* 108 no. 4 (1978): 312–21, https://doi.org/10.1093/oxfordjournals.aje.a112625.

88 Not long after: Shana Elbaum-Garfinkle, "Close to Home: A History of Yale and Lyme Disease," *Yale Journal of Biology Medicine* 84, no. 2 (2011): 103–8, https://www.ncbi.nlm.nih.gov/pmc/articles/PMC3117402/.

88 central nervous system: Elbaum-Garfinkle, "Close to Home."

88 skin: Marjorie Hecht, "13 Signs and Symptoms of Lyme Disease," Healthline, May 16, 2018, updated on March 29, 2019, https://www.healthline.com/health/lyme-disease-symptoms#skin-outbreaks.

89 people are traveling: Nina Bai, "Lyme Disease Is on the Rise—An Expert Explains Why," University of California, San Francisco, Research, May 16, 2018, https://www.ucsf.edu/news/2018/05/410401/lyme-disease-rise-expert-explains-why.

90 urbanization: "How Urbanization Affects Spread of Lyme Disease," Infection Control Today, June 27, 2019, https://www.infectioncontroltoday.com/infectious-diseases-conditions/how-urbanization-affects-spread-lyme-disease.

90 the number of tick-borne: R. Rosenberg et al., "Vital Signs: Trends in Reported Vectorborne Disease Cases—United States and Territories, 2004–2016," CDC, *Morbidity and Mortality Weekly Report* 67 (2018): 496–501, last reviewed May 3, 2018, https://www.cdc.gov/mmwr/volumes/67/wr/mm6717e1.htm.

90 Around 300,000: "CDC Provides Estimate of Americans Diagnosed with Lyme Disease Each Year," CDC Newsroom, August 19, 2013, https://www.cdc.gov/media/releases/2013/p0819-lyme-disease.html.

90 That's over 1.5 times: L. Johnson et al., "Severity of Chronic Lyme Disease Compared to Other Chronic Conditions: A Quality of Life Survey," *PeerJ* 2, no. 322 (March 2014), https://doi.org/10.7717/peerj.322.

90 By 2018: CBS News, "Lyme Disease Has Now Spread to All 50 States, Report Finds," August 7, 2018, https://www.cbsnews.com/news/lyme-disease-has-now-spread-to-all-50-states-report-finds/.

96 10 to 20 percent: Michael T. Melia and Paul G. Auwaerter, "Time for a Different Approach to Lyme Disease and Long-Term Symptoms," *New England Journal of Medicine* 374 (2016): 1277–78, https://doi.org/10.1056/NEJMe1502350.

96 the only one recommended: CDC, "Lyme Disease: Diagnosis and Testing," Centers for Disease Control and Prevention, last reviewed November 20, 2019, https://www.cdc.gov/lyme/diagnosistesting/index.html.

98 One of the four founders: "The Founding Physicians," About Johns Hopkins Medicine, hopkinsmedicine.org, https://www.hopkinsmedicine.org/about/history/history5.html.

98 almost every psychiatric disorder: L. L. Crozatti et al., "Atypical Behavioral and Psychiatric Symptoms: Neurosyphilis Should Always Be Considered," *Autopsy Case Reports* 5, no. 3 (September 2015): 43–47, https://doi.org/10.4322/acr.2015.021.

99 "Know syphilis": W. B. Bean, *Sir William Osler: Aphorisms from His Bedside Teachings and Writings* (Springfield, IL: Charles C. Thomas, 1968), 133.

102 as many as 80 percent: CDC, "Symptoms of Tickborne Illness," last reviewed January 10, 2019, https://www.cdc.gov/ticks/symptoms.html.

102 others saying: A. Stonehouse, J. S. Studdiford, C. A. Henry, "An Update on the Diagnosis and Treatment of Early Lyme Disease: 'Focusing on the Bull's Eye, You May Miss the Mark,' Journal of Emergency Medicine 39, no. 5 (2010): e147–51, https://doi.org/10.1016/j.jemermed.2007.06.007; Lorraine Johnson, "How Many of Those With Lyme Disease Have the Rash? Estimates Range From 27–80%," Lyme Policy Wonk, April 10, 2014, https://www.lymedisease.org/lymepolicywonk-how-many-of-those-with-lyme-disease-have-the-rash-estimates-range-from-27-80-2/.

102 about 50 percent: John O. Meyerhoff, Russell W. Steele, and Gerald W. Zaidman, "Lyme Disease Clinical Presentation," Medscape, updated August 1, 2019, https://emedicine.medscape.com/article/330178-clinical.

103 In late-stage Lyme: Medical Knowledge Team, Ada Health, "Late Lyme Disease," updated February 18, 2020, https://ada.com/conditions/late-lyme-disease/#symptoms.

105 change form: Lynn Margulis et al., "Spirochete Round Bodies Syphilis, Lyme Disease & AIDS: Resurgence of 'The Great Imitator'?," *Symbiosis* 47 (2009): 51–58, https://doi.org/10.1007/BF03179970.

105 when *Borrelia burgdorferi* morph: L. Meriläinen et al., "Morphological and Biochemical Features of *Borrelia burgdorferi* Pleomorphic Forms," *Microbiology* 161, no. 3 (2015): 516–27, https://doi.org/10.1099/mic.0.000027.

105 "persisters": Eva Sapi et al., "The Long-Term Persistence of *Borrelia burgdorferi* Antigens and DNA in the Tissues of a Patient with Lyme Disease," *Antibiotics* 8, no. 4 (October 2019): 183, https://doi.org.10.3390/antibiotics8040183.

105 Once the environment: Margulis et al., "Spirochete Round Bodies," 51–58.

107 Homicidal thoughts: Robert C. Bransfield, "Aggressiveness, Violence, Homicidality, Homicide, and Lyme Disease," *Neuropsychiatric Disease and Treatment* 14, no. 9 (March 2018): 693–713, doi:10.2147/NDT.S155143.

107 slightly over 50 percent: Johnson, et al., "Severity of Chronic Lyme Disease."

108 intense, violent anger: Robert C. Bransfield, "Suicide and Lyme and Associated Diseases," *Neuropsychiatric Disease and Treatment* 13 (2017): 1575–87,

http://doi.org/10.2147/NDT.S136137; Edward B. Breitschwerdt et al., "*Bar-tonella henselae* Bloodstream Infection in a Boy with Pediatric Acute-Onset Neuropsychiatric Syndrome," *Journal of Central Nervous System Disease* 11 (March 2019): 1–8, https://doi.org/10.1177/1179573519832014.

108 kill the infection: S. K. Singh and H. J. Girschick, "Toll-like Receptors in *Borrelia burgdorferi*-Induced Inflammation," *Clinical Microbiology and Infection* 12, no. 8 (August 2006): 705–17, https://doi.org/10.1111/j.1469-0691 .2006.01440.x.

109 about half of patients: Abhishek Chandra et al., "Anti-Neural Antibody Re-activity in Patients with a History of Lyme Borreliosis and Persistent Symp-toms," *Brain, Behavior, and Immunity* 24, no. 6 (August 2010): 1018–24, https://doi.org/10.1016/j.bbi.2010.03.002.

109 emerging links: Matthew S. Kayser and Josep Dalmau, "The Emerging Link Between Autoimmune Disorders and Neuropsychiatric Disease," *Journal of Neuropsychiatry and Clinical Neurosciences,* 23, no.a (Fall 2011): 90–97, https://doi.org/10.1176/appi.neuropsych.23.1.90.

110 which are more sensitive: "IGeneX Inc. Introduces New Diagnostic Tests for Lyme Disease and Tick-Borne Relapsing Fever," IGeneX Inc. press release, https://igenex.com/press-release/igenex-inc-introduces-new-diagnostic-tests -for-lyme-disease-and-tick-borne-relapsing-fever/.

113 an amino acid that has shown: Flavia di Michele et al., "N-Acetyl Cyste-ine and Vitamin D Supplementation in Treatment Resistant Obsessive-Compulsive Disorder Patients: A General Review," *Current Pharmaceutical Design* 24, no. 17 (2018):1832–38, https://doi.org/10.2174/138161282466 6180417124919; Georgina Oliver et al., "N-Acetyl Cysteine in the Treatment of Obsessive Compulsive and Related Disorders: A Systematic Review," *Clin-ical Psychopharmacology and Neuroscience* 13, no. 1 (2015): 12–24, https://doi .org/10.9758/cpn.2015.13.1.12.

116 2017 Lyme Disease case: National Notifiable Diseases Surveillance System (NNDSS), "Lyme Disease (*Borrelia burgdorferi*) 2017 Case Definition," Centers for Disease Control, https://wwwn.cdc.gov/nndss/conditions/lyme -disease/case-definition/2017/.

CHAPTER 6: STOPPED COLD

121 via the bloodstream: Robert M. Sargis, "How Your Thyroid Works," Endo-crineweb, updated October 21, 2019, https://www.endocrineweb.com/conditions /thyroid/how-your-thyroid-works.

121 muscle contraction: Healthwise Staff, "Thyroid Hormone Production and Function," University of Michigan, Michigan Medicine Health Library, current as of July 28, 2019, https://www.uofmhealth.org/health-library/ug1836.

121 the active components: L. Bartalena and J. Robbins, "Thyroid Hormone Transport Proteins," *Clinics in Laboratory Medicine* 13, no. 3 (September 1993): 583–598, https://www.ncbi.nlm.nih.gov/pubmed/8222576.

121 brain, liver, and kidneys: Healthbeat, "Could It Be My Thyroid?" Harvard Health Publishing, https://www.health.harvard.edu/healthbeat/could-it-be-my -thyroid.

122 When the thyroid: Stefano Mariotti and Paolo Beck-Peccoz, "Physiology of the Hypothalamic-Pituitary-Thyroid Axis," Endotext, last updated August 14, 2016, https://www.ncbi.nlm.nih.gov/books/NBK278958/.

122 According to: American Thyroid Association, "Prevalence and Impact of Thyroid Disease," General Information/Press Room, https://www.thyroid .org/media-main/press-room/.

122 About 60 percent of those people: American Thyroid Association, "Prevalence and Impact."

122 second only to diabetes: J. M. Kuyl, "The Evolution of Thyroid Function Tests," *Journal of Endocrinology, Metabolism and Diabetes of South Africa* 20, no. 2 (2015), http://www.jemdsa.co.za/index.php/JEMDSA/article/view/490.

123 Until the early twentieth century: Elizabeth A. McAninch and Antonio C. Bianco, "The History and Future of Treatment of Hypothyroidism," *Annals of Internal Medicine* 164, no. 1 (2016): 50–56. https://doi.org/10.7326/M15-1799.

123 500 BCE: Luigi Massimino Sena, "The Thyroid in Art," https://dn3g20un7g odm.cloudfront.net/2011/AM11SA/161.pdf.

123 Adult patients: Thomas W. Heinrich and Garth Grahm, "Hypothyroidism Presenting as Psychosis: Myxedema Madness Revisited," *Primary Care Companion to the Journal of Clinical Psychiatry* 5, no. 6 (2003): 260–66, https:// doi.org/10.4088/pcc.v05n0603.

123 after burning himself with acid: N. Zorkin et al., "Severe Hypothyroidism Presenting with Acute Mania and Psychosis: A Case Report and Literature Review," *Bipolar Disorder* 3, no. 1 (2017), https://doi.org/10.4172/2472 -1077.1000116.

123 1 to 4 percent: Anna Z. Feldman, Rupendra T. Shrestha, and James V. Hennessey, "Neuropsychiatric Manifestations of Thyroid Disease," *Endocrinology and Metabolism Clinics of North America* 42 (2013): 453–76, https://doi .org/10.1016/j.ecl.2013.05.005.

123 this particular patient: Zorkin et al., "Severe Hypothyroidism."

123 in some cases: R. Asher, "Myxoedematous Madness," *British Medical Journal* (September 10, 1949): 555–62, https://www.ncbi.nlm.nih.gov/pmc/articles /PMC2051123/pdf/brmedj03641-0005.pdf.

123 five to eight times: American Thyroid Association, "Prevalence and Impact."

124 The risk generally rises with age: Heinrich and Grahm, "Hypothyroidism Presenting as Psychosis," 260–66.

124 two to three of every hundred children: Sari Harrar, "5 Surprising Facts about 'Low Thyroid' in Children and Teens," Endocrineweb, updated on January 19, 2017, https://www.endocrineweb.com/conditions/hypothyroidism/5-surprising -facts-about-low-thyroid-children-teens.

124 20 million people: American Thyroid Association, "Prevalence and Impact."

124 14 million have Hashimoto's: Kresimira Milas, "Hashimoto's Thyroiditis Facts and Tips," Endocrineweb, last updated May 29, 2014, https://www.endocrineweb .com/conditions/hashimotos-thyroiditis/hashimotos-thyroiditis-facts-tips.

124 a known trigger: Yhojan Rodríguez et al., "Tick-Borne Diseases and Auto-immunity: A Comprehensive Review," *Journal of Autoimmunity* 88 (March 2018): 21–42, https://doi.org/10.1016/j.jaut.2017.11.007.

126 The degree of thyroid dysfunction: Heinrich and Grahm, "Hypothyroidism Presenting as Psychosis," 260–66.

126 Hashimoto's symptoms can precede: Amy Myers, "3 Reasons Your Doctor Missed Your Hashimoto's Diagnosis," Amy Myers MD, July 1, 2020, https://www .amymyersmd.com/2018/02/3-reasons-doctor-missed-hashimotos-diagnosis/.

129 the chemicals in our environment: Shan Juan, "Thyroid Diseases on the Rise in China: Expert," ChinaDaily.com, last updated June 3, 2016, http://www .chinadaily.com.cn/china/2016-06/03/content_25607805.htm.

130 Some studies have shown: Valeria Calsolaro, et al., "Thyroid Disrupting Chemicals," *International Journal of Molecular Sciences* 18, no. 12 (December 2017): 2583, https://doi.org/10.3390/ijms18122583.

130 banned in the late 1970s: R. Thomas Zoeller et al., "Endocrine-Disrupting Chemicals and Public Health Protection: A Statement of Principles from the Endocrine Society," *Endocrinology* 153, no. 9 (2012): 4097–110, https://doi .org/10.1210/en.2012-1422.

130 Dioxins: "Dioxins and Their Effects on Human Health," World Health Organization, who.int, October 4, 2016, https://www.who.int/news-room/fact -sheets/detail/dioxins-and-their-effects-on-human-health.

130 increase TSH levels: M. M. Leijs et al., "Thyroid Hormone Metabolism and Environmental Chemical Exposure," *Environmental Health* 11, no. 1 (2012), https://doi.org/10.1186/1476-069X-11-S1-S10.

130 a patient can have labs: Leijs et al., "Thyroid Hormone."

130 Perchlorate: Roddy Scheer and Doug Moss, "Perchlorate in Drinking Water Raises Health Concerns," *E-The Environmental Magazine, Scientific American,* December 21, 2012, https://www.scientificamerican.com/article/perchlorate -in-drinking-water/.

130 known to bind: Erik Stokstad, "Perchlorate Impacts Thyroid at Low Doses," *Science,* October 6, 2006, https://www.sciencemag.org/news/2006/10/perchlorate -impacts-thyroid-low-doses; Giuseppe Lisco, et al., "Interference on Iodine Uptake and Human Thyroid Function by Perchlorate-Contaminated Water and Food," *Nutrients* 12, no. 6 (2020): 1669, https://doi.org/10.3390/nu 12061669.

130 public drinking water: Scheer and Moss, "Perchlorate in Drinking Water"; Rebecca Renner, "Perchlorate in Food," *Environmental Science and Technology,* March 15, 2008, 1817, https://pubs.acs.org/doi/pdf/10.1021/es0870552.

130 Fluoride: Navneet Singh et al., "A Comparative Study of Fluoride Ingestion Levels, Serum Thyroid Hormone & TSH Level Derangements, Dental Fluorosis Status among School Children from Endemic and Non-Endemic Fluorosis Areas," *SpringerPlus* 3, no. 7 (January 3, 2014), doi:10.1186/2193 -1801-3-7.

131 food sensitivities: M. A. Iddah and B. N. Macharia, "Autoimmune Thyroid Disorders," *International Scholarly Research Notices Endocrinology* 2013, article 509764 (June 2013), https://doi.org/10.1155/2013/509764; Rui Lin et al., "Abnormal Intestinal Permeability and Microbiota in Patients with Autoimmune Hepatitis," *International Journal of Clinical And Experimental Pathology* 8, no. 5 (May 2015): 5153–60.

131 Increased intestinal permeability: Qinghui Mu et al., "Leaky Gut as a Danger Signal for Autoimmune Diseases," *Frontiers In Immunology* 8, no. 23 (May 2017), doi:10.3389/fimmu.2017.00598.

131 15 percent of childhood: Patrick Hanley, Katherine Lord, and Andrew J. Bauer, "Thyroid Disorders in Children and Adolescents: A Review," *JAMA Pediatrics* 170, no. 10 (October 2016): 1014, https://www.sfedp.org /wp-content/uploads/2017/06/Review-sur-la-pathologie-thyroidienne-p %C3%A9diatrique-JAMA-Ped-2016.pdf.

131 Physical symptoms: Hanley, et al., "Thyroid Disorders in Children," 1013.

131 overactive reflexes: "Hyperthyroidism (Overactive Thyroid)," Mayo Clinic, Diseases and Conditions, Mayoclinic.org, updated January 7, 2020, https:// www.mayoclinic.org/diseases-conditions/hyperthyroidism/diagnosis -treatment/drc-20373665.

131 sensitivity to heat: Healthbeat, "Could It Be My Thyroid?"

131 Puberty: Hanley et al., "Thyroid Disorders in Children," 1012.

132 sometimes confused with: Hanley et al., "Thyroid Disorders in Children," 1014.

132 isn't because it's natural: Harrar, "5 Surprising Facts."

132 there have been cases: Sarah Kotchen Rock and Eugene H. Makela, "Hypothyroidism and Depression: A Therapeutic Challenge," *Annals of Pharmacotherapy* 34, no. 10, (October 2000), https://doi.org/10.1345/aph.10022.

133 vitamin D: Robert Krysiak, Witold Szkróbka, and Boguslaw Okopień, "The Effect of Vitamin D and Selenomethionine on Thyroid Antibody Titers, Hypothalamic-Pituitary-Thyroid Axis Activity and Thyroid Function Tests in Men with Hashimoto's Thyroiditis: A Pilot Study," *Pharmacological Reports* 71, no. 2 (October 2018): 243–47, https://doi.org/10.1016/j.pharep.2018.10.012.

133 Researchers have been working on: Michael I. Liontiris and Elias E. Mazokopakis, "A Concise Review of Hashimoto Thyroiditis (HT) and the Importance of Iodine, Selenium, Vitamin D and Gluten on the Autoimmunity and Dietary Management of HT Patients. Points That Need More Investigation," *Hellenic Journal of Nuclear Medicine* 20, no. 1 (2017): 51–56, http://shayahealth.com/resources/Hashimotos%20gluten%20review%20article.pdf; George J. Kahaly, Lara Frommer, and Detlef Schuppan, "Celiac Disease and Glandular Autoimmunity," *Nutrients* 814, no. 10 (2018): doi: 10.3390/nu10070814.

133 a more recent study: Robert Krysiak, Witold Szkróbka, and Boguslaw Okopień, "The Effect of Gluten-Free Diet on Thyroid Autoimmunity in Drug-Naïve Women with Hashimoto's Thyroiditis: A Pilot Study," *Experimental and Clinical Endocrinology & Diabetes* 127, no. 7 (July 2019): 417–22, https://doi.org/10.1055/a-0653-7108.

CHAPTER 7: DEAD ON THEIR FEET

140 according to some studies: Lisa Rapaport, "Healthy Kids with Sick Sibling May Hide Emotions," Reuters Health News, July 26, 2018, https://www.reuters.com/article/us-health-kids-ill-sibling/healthy-kids-with-sick-sibling-may-hide-emotions-idUSKBN1KG31A.

142 it's not uncommon: Kristin H. Lagattuta, L. Sayfan, and C. Bamford, "Do You Know How I Feel? Parents Underestimate Worry and Overestimate Optimism Compared to Child Self-Report," *Journal of Experimental Child Psychology* 113, no. 2 (2012): 211–32, https://www.sciencedirect.com/science/article/abs/pii/S0022096512000616.

147 salivary cortisol tests: Christina S. Chao et al., "Salivary Cortisol Levels by Tandem Mass Spectrometry During High Dose ACTH Stimulation Test for Adrenal Insufficiency in Children," *Endocrine* 67, no. 1 (January 2019), https://doi.org/10.1007/s12020-019-02084-8; M. H. Laudat et al, "Salivary Cortisol Measurement: A Practical Approach to Assess Pituitary-Adrenal Function," *Journal of Clinical Endocrinology & Metabolism* 66, no. 2 (February 1988): 343–48, https://doi.org/10.1210/jcem-66-2-343.

148 many children: John D. Guerry and Paul D. Hastings, "In Search of HPA Axis Dysregulation in Child and Adolescent Depression," *Clinical Child and Family Psychology Review* 14, no. 2 (2011): 135–60, https://doi.org/10.1007/s10567-011-0084-5.

148 helps shape our stress response: Patrick O. McGowan, "Epigenomic Mechanisms of Early Adversity and HPA Dysfunction: Considerations for PTSD Research," *Frontiers in Psychiatry* 4, no. 110 (September 2013), https://doi.org/10.3389/fpsyt.2013.00110.

148 bipolar disorder: M. Belvederi Murri et al., "The HPA Axis in Bipolar Disorder: Systematic Review and Meta-Analysis," *Psychoneuroendocrinology* 63 (2016): 327–42, https://doi.org/10.1016/j.psyneuen.2015.10.014.

148 depression: McGowan, "Epigenomic Mechanisms"; Guerry and Hastings, "In Search of HPA Axis Dysregulation."

148 sustained stress: Amanda Brown et al., "Bridging Basic and Clinical Research in Early Life Adversity, DNA Methylation, and Major Depressive Disorder," *Frontiers in Genetics* 10, no. 229 (March 2019), https://doi.org/10.3389/fgene.2019.00229.

148 these regulatory systems can be repaired: Jayanthi Maniam et al., "Early-Life Stress, HPA Axis Adaptation, and Mechanisms Contributing to Later Health Outcomes," *Frontiers in Endocrinology* 5, no. 73 (May 2014): https://doi.org/10.3389/fendo.2014.00073.

149 Studies have shown a connection between a high glycemic index: James E. Gangwisch et al., "High Glycemic Index Diet as a Risk Factor for Depression: Analyses from the Women's Health Initiative," *American Journal of Clinical Nutrition* 102, no. 2 (2015): 454–63, https://doi.org/10.3945/ajcn.114.103846; Liam Davenport, "Role of Nutrition in Mental Health a Public Health Matter," Medscape, July 3, 2020, https://www.medscape.com/viewarticle/933386#vp_2; Ye Li et al., "Dietary Patterns and Depression Risk: A Meta-Analysis," *Psychiatry Research* 253 (July 2017): 373–82, https://doi.org/10.1016/j.psychres.2017.04.020.

149 some studies have shown that switching: Gangwisch et al., "High Glycemic

Index Diet"; Maria Cohut, "Mediterranean Diet May Protect Against Depression Symptoms," Medical News Today, May 22, 2019, https://www.medicalnewstoday.com/articles/325240; Heather M. Francis et al., "A Brief Diet Intervention Can Reduce Symptoms of Depression in Young Adults—A Randomised Controlled Trial," *PLoS ONE* 14, no. 10 (October 2019): e0222768, https://doi.org/10.1371/journal.pone.0222768; Monique Aucoin and Sukriti Bhardwaj, "Generalized Anxiety Disorder and Hypoglycemia Symptoms Improved with Diet Modification," *Case Reports in Psychiatry* 2016, no. 7165425 (2016), https://doi.org/10.1155/2016/7165425.

150 Around 90 percent: Gita Shafiee et al., "The Importance of Hypoglycemia in Diabetic Patients," *Journal of Diabetes and Metabolic Disorders* 11, no. 1 (October 2012), doi:10.1186/2251-6581-11-17.

150 Reactive hypoglycemia: Harry M. Salzer, "Relative Hypoglycemia as a Cause of Neuropsychiatric Illness," *Journal of the National Medical Association* 58, no.1 (January 1966): 12, PMCID: PMC2611193.

150 both functional hypoadrenalism and reactive hypoglycemia are conditions that many doctors claim don't exist: S. Chalew et al., "The Use of the Plasma Epinephrine Response in the Diagnosis of Idiopathic Postprandial Syndrome," *Journal of the American Medical Association* 251 (February 1984): 612–15, https://doi.org/10.1001/jama.1984.03340290026014.

157 the adrenal glands release: Eileen K. Corrigan, "Adrenal Insufficiency (Addison's Disease)," Pituitary Network Association, https://pituitary.org/knowledge-base/disorders/adrenal-insuffieciency-addison-s-disease.

159 high glycemic eating habits: A. A. Kulkarni, B. A. Swinburn, and J. Utter, "Associations Between Diet Quality and Mental Health in Socially Disadvantaged New Zealand Adolescents," *European Journal of Clinical Nutrition* 69, no. 1 (2015): 79–83, https://doi.org/10.1038/ejcn.2014.130; Gangwisch et al., "High Glycemic Index Diet"; F. Haghighatdoost et al., "Glycemic Index, Glycemic Load, and Common Psychological Disorders," *American Journal of Clinical Nutrition* 103, no. 1 (2016): 201–9, https://doi.org/10.3945/ajcn.114.105445; R. A. Cheatham et al., "Long-Term Effects of Provided Low and High Glycemic Load Low Energy Diets on Mood and Cognition," *Physiology and Behavior* 98 no. 3 (2009): 374–79, https://doi.org/10.1016/j.physbeh.2009.06.015; M. Hosseinzadeh et al., "Empirically Derived Dietary Patterns in Relation to Psychological Disorders," *Public Health Nutrition* 19 no. 2 (February 2016): 204–17, doi: 10.1017/S136898001500172X.

159 a fifteen-year-old: Aucoin and Bhardwaj, "Generalized Anxiety Disorder."

CHAPTER 8: OVERREACTING TO LIFE

165 more than 50 million: "Allergy Facts and Figures," Asthma and Allergy Foundation of America, https://www.aafa.org/allergy-facts/.

165 it's estimated that 5.6 million children: "Food Allergy Facts and Statistics for the U.S.," Food Allergy Research and Education, last reviewed June 8, 2020, https://www.foodallergy.org/life-with-food-allergies/food-allergy-101/facts -and-statistics.

166 About 90 percent: "Food Allergies in Children," Johns Hopkins Medicine .org, https://www.hopkinsmedicine.org/health/conditions-and-diseases/food -allergies-in-children.

168 The FDA-enforced Food Allergen Labeling and Consumer Protection Act: U.S. Food and Drug Administration, "What You Need to Know About Food Allergies," FDA.gov, content current as of September 26, 2018, https://www .fda.gov/food/buy-store-serve-safe-food/what-you-need-know-about-food -allergies.

168 "pure" vanilla: Erica Marcus, "A Tweak of Sweet in Vanilla Extract," Newsday .com, March 2, 2011, https://www.newsday.com/lifestyle/restaurants/a-tweak -of-sweet-in-vanilla-extract-1.2722815.

168 is made with 35 percent ethyl alcohol: Caroline Stanko, "Does Vanilla Extract Contain Alcohol?," *Taste of Home*, https://www.tasteofhome.com/article/alcohol -in-extracts/.

168 can be derived from corn: The Editors of Encyclopaedia Britannica, "Ethanol," Britannica.com, Conditions and Diseases, last updated March 28, 2019, https://www.britannica.com/science/ethanol.

168 toothpaste, makeup: "12 Surprising Products That Contain Corn," Forbes. com, https://www.forbes.com/pictures/fejg45geil/12-surprising-products-that -contain-corn/#6d2c9d215bd8.

169 sometimes symptoms that appear: Joshua D. Rosenblat et al., "Inflamed Moods: A Review of the Interactions Between Inflammation and Mood Disorders," *Progress in Neuro-Psychopharmacology & Biological Psychiatry*, 53 (2014): 23–34, https://doi.org/10.1016/j.pnpbp.2014.01.013.

169 One population-based study: N.-S. Tzeng et al., "Increased Risk of Psychiatric Disorders in Allergic Diseases: A Nationwide, Population-Based, Cohort Study," *Frontiers in Psychiatry* 9, no. 133 (April 2018), https://doi .org/10.3389/fpsyt.2018.00133.

169 Cow's milk: Nicholas A. Smith et al., "Astrogliosis Associated with Behavioral Abnormality in a Non-anaphylactic Mouse Model of Cow's Milk Allergy,"

Frontiers in Cellular Neuroscience 13, no. 320 (July 2019): https://doi.org /10.3389/fncel.2019.00320.

170 doctors have seen positive results: Gordon Parker and Tim Watkins, "Treatment-Resistant Depression: When Antidepressant Drug Intolerance May Indicate Food Intolerance," *Australian and New Zealand Journal of Psychiatry* 36 no. 2, (2002): 263–65, 10.1046/j.1440-1614.2002.00978.x.

170 Consistently, elevated levels of IL-6: Jeffrey L. Voorhees, et al., "Prolonged Restraint Stress Increases IL-6, Reduces IL-10, and Causes Persistent Depressive-Like Behavior That Is Reversed by Recombinant IL-10," *PLOS One* (March 8, 2013), https://doi.org/10.1371/journal.pone.0058488.

170 and anxiety: Ana F. Trueba, Thomas Ritz, and Gabriel Trueba, "The Role of the Microbiome in the Relationship of Asthma and Affective Disorders," *Microbial Endocrinology: Interkingdom Signaling in Infectious Disease and Health. Advances in Experimental Medicine and Biology* 874 (2016): 263–88, https://doi.org/10.1007/978-3-319-20215-0_13.

170 and allergies: Y. Lin, S. Chen, and J. Wang, "Critical Role of IL-6 in Dendritic Cell-induced Allergic Inflammation of Asthma" *Journal of Molecular Medicine,* 94 (2016): 51–59, https://doi.org/10.1007/s00109-015-1325-8; Trueba, Ritz, and Trueba, "The Role of the Microbiome," 263–288. L P. Gosset, F. Malaquin, Y. Delneste, et.al., "Interleukin-6 and Interleukin-1 Alpha Production Is Associated with Antigen-Induced Late Nasal Response," *Journal of Allergy and Clinical Immunology* 92, no. 6 (1993): 878–90, https://doi .org/10.1016/0091-6749(93)90066-o.

170 According to the CDC: Food Allergy and Anaphylaxis Connection Team, "Food Allergy Basics," Food AllergyAwareness.org, https://www.foodallergy awareness.org/food-allergy-and-anaphylaxis/food-allergy-basics/food-allergy -basics/.

170 No one is really sure why: Alexandra Santos, "Why the World Is Becoming More Allergic to Food," BBC News, September 13, 2019, https://www.bbc .com/news/health-46302780.

170 offering allergenic foods: Jane L. Holl et al., "A Randomized Trial of the Acceptability of a Daily Multi-Allergen Food Supplement for Infants," *Pediatric Allergy and Immunology* 31, no. 4 (February 2020), https://doi.org/10.1111/pai.13223.

170 overreliance on antibacterial: Santos, "Why the World."

170 obesity: Sara Benedé et al., "The Rise of Food Allergy: Environmental Factors and Emerging Treatments," *EBioMedicine* 7 (May 2016): 27–34. https://doi .org/10.1016/j.ebiom.2016.04.012.

170 vitamin D deficiency: Santos, "Why the World."

170 loss of the integrity: Nazanin Samadi, Martina Klems, and Eva Untersmayr, "The Role of Gastrointestinal Permeability in Food Allergy," *Annals of Allergy, Asthma, and Immunology* 121, no. 2 (August 2018): 166–73, https://doi.org/10.1016/j.anai.2018.05.010.

171 only 1 percent: C. Perrier and B. Corthésy, "Gut Permeability and Food Allergies," *Clinical and Experimental Allergy* 41, no. 1 (2011): 20–28, https://doi.org/10.1111/j.1365-2222.2010.03639.x.

173 a term that's no longer: "What Is PDD-NOS?," Healthline.com, last reviewed April 29, 2019, https://www.healthline.com/health/autism/pdd-nos#pdd-nos.

175 it really means: Elana Lavine, "Blood Testing for Sensitivity, Allergy or Intolerance to Food," *Canadian Medical Association Journal* 184, no. 6 (2012): 666–68, https://doi.org/10.1503/cmaj.110026.

176 gluten can have a direct effect: Eleanor Busby et al., "Mood Disorders and Gluten: It's Not All in Your Mind! A Systematic Review with Meta-Analysis," *Nutrients* 10, no. 11 (November 2018): 1708, https://doi.org/10.3390/nu10111708; Marios Hadjivassiliou et al., "Gluten Sensitivity: From Gut to Brain," *Lancet* 9, no. 3 (March 2010): 318–30, https://doi.org/10.1016/S1474-4422(09)70290-X; S. L. Peters et al., "Randomised Clinical Trial: Gluten May Cause Depression in Subjects with Non-Coeliac Gluten Sensitivity: An Exploratory Clinical Study," *Alimentary Pharmacology & Therapeutics* 39, no. 10 (2014): 1104–12, https://doi.org/10.1111/apt.12730.

176 gluten has properties: Leo Pruimboom and Karin de Punder, "The Opioid Effects of Gluten Exorphins: Asymptomatic Celiac Disease," *Journal of Health, Population, and Nutrition* 33, no. 24 (November 2015), https://doi.org/10.1186/s41043-015-0032-y.

176 bidirectional nature of the gut-brain-immune: Hadjivassiliou et al., "Gluten Sensitivity."

176 that inflammation can promote gut dysbiosis: Mak Adam Daulatzai, "Non-Celiac Gluten Sensitivity Triggers Gut Dysbiosis, Neuroinflammation, Gut-Brain Axis Dysfunction, and Vulnerability for Dementia," *CNS & Neurologial Disorders: Drug Targets* 14, no. 1 (2015): 110–31, https://doi.org/10.2174/1871527314666150202152436.

176 affecting mood: Therese Borchard, "Gluten, Depression, and Anxiety: The Gut-Brain Link," Everydayhealth.com, September 1, 2016, https://www.everydayhealth.com/columns/therese-borchard-sanity-break/gluten-depression-and-anxiety-gut-brain-link/; Megan Clapp et al., "Gut Microbiota's Effect on Mental Health: The Gut-Brain Axis," *Clinics and Practice* 7, no. 4 (September 2017): 987, https://doi.org/10.4081/cp.2017.987.

176 As far back as the 1950s: Guy Daynes, "Bread and Tears—Naughtiness, Depression, and Fits Due to Wheat Sensitivity," *Proceedings of the Royal Society of Medicine* 49 (February 1956): 19, https://www.ncbi.nlm.nih.gov/pmc/articles/PMC1889136/pdf/procrsmed00381-0033.pdf.

177 In 1966, researchers found a link: F. C. Dohan, "Wheat 'Consumption' and Hospital Admissions for Schizophrenia During World War II: A Preliminary Report," *American Journal of Clinical Nutrition* 18, no. 1 (January 1966): 7–10, https://doi.org/10.1093/ajcn/18.1.7.

177 In 2012, a girl: Elena Lionetti et al., "Gluten Psychosis: Confirmation of a New Clinical Entity," *Nutrients* 7, no. 7 (July 2015): 5532–39, https://doi.org/10.3390/nu7075235.

177 test negative for both celiac disease: Pasquale Mansueto et al., "Non-Celiac Gluten Sensitivity: Literature Review," *Journal of the American College of Nutrition* 33, no. 1 (2014): 39–54, doi:10.1080/07315724.2014.869996.

177 frequently test positive: Jessica R. Jackson et al., "Neurologic and Psychiatric Manifestations of Celiac Disease and Gluten Sensitivity," *Psychiatric Quarterly* 83, no. 1 (2012): 91–102, https://doi.org/10.1007/s11126-011-9186-y.

177 symptoms are not limited: Maria Raffaella Barbaro et al., "Recent Advances in Understanding Non-Celiac Gluten Sensitivity," *F1000Research* 7 (October 2018), https://doi.org/10.12688/f1000research.15849.1.

177 and depression: Antonio Carroccio et al., "Self-Reported Non-Celiac Wheat Sensitivity in High School Students: Demographic and Clinical Characteristics," *Nutrients* 9, no. 7 (July 2017): 771, https://doi.org/10.3390/nu9070771.

178 The dramatic mental: Busby et al., "Mood Disorders," 1708.

181 "convergence insufficiency": Mayo Clinic Staff, "Convergence Insufficiency," July 15, 2017, Mayoclinic.org, Diseases and Conditions, https://www.mayoclinic.org/diseases-conditions/convergence-insufficiency/symptoms-causes/syc-20352735.

181 affecting depth perception: Carmen Willings, "Convergence Insufficiency," Teaching Students with Visual Impairments, teachingvisuallyimpaired.com, https://www.teachingvisuallyimpaired.com/convergence-insufficiency-ci.html.

182 a prestigious medical journal: Donna McCann et al., "Food Additives and Hyperactive Behaviour in 3-Year-Old and 8/9-Year-Old Children in the Community: A Randomised, Double-Blinded, Placebo-Controlled Trial," *Lancet* 370, no. 9598 (November 2007): 1560–67, https://doi.org/10.1016/S0140-6736(07)61306-3.

184 found in many foods: Kelley Reed, "List of High Phenol Foods," Livestrong.com, https://www.livestrong.com/article/165497-list-of-high-phenol-foods/.

186 combat stage fright: James A. Bourgeois, "The Management of Performance Anxiety with Beta-Adrenergic Blocking Agents," *Jefferson Journal of Psychiatry* 9, no. 2, article 5 (June 1991), https://doi.org/10.29046/JJP.009.2.002.

186 particularly those with Asperger's: K. Sagar-Ouriaghli et al., "Effectiveness of Propranolol for Treating Anxiety and Aggression in Children and Adolescents with Autism Spectrum Disorder," *Annual Meeting of the International Society for Autism Research*, May 12, 2017.

186 it can work well: Richard Famularo, R. Kinscherff, and T. Fenton, "Propranolol Treatment for Childhood Posttraumatic Stress Disorder, Acute Type. A Pilot Study," *American Journal of Diseases of Children* 142, no. 11 (1988): 1244–1247, https://doi.org/10.1001/archpedi.1988.02150110122036.

CHAPTER 9: BOTTLENECKED

190 scurvy: Simon Worrall, "A Nightmare Disease Haunted Ships During Age of Discovery," *National Geographic*, January 15, 2017, https://www.national geographic.com/news/2017/01/scurvy-disease-discovery-jonathan-lamb/.

190 rickets: University of Bristol, "Study Identifies Prevalence of Rickets Among 16th Century Sailors," Phys.org, December 17, 2014, https://phys.org/news /2014-12-prevalence-rickets-16th-century-sailors.html.

190 a diet low in: Kebashni Thandrayen and John M. Pettifor, "The Roles of Vitamin D and Dietary Calcium in Nutritional Rickets," *Bone Reports* 8 (June 2018): 81–89, 10.1016/j.bonr.2018.01.005.

190 Low magnesium levels: Temma Ehrenfeld, "Magnesium Might Boost Mood," *Psychology Today*, September 2107, https://www.psychologytoday.com/us /blog/open-gently/201709/magnesium-might-boost-mood.

191 leave the brain primed: Lin Wan et al. "Methylenetetrahydrofolate Reductase and Psychiatric Diseases," *Translational Psychiatry* 8, no. 1 (November 2018): 242, https://doi.org/10.1038/s41398-018-0276-6.

191 up to 70 percent of patients with depression: Richard C. Shelton et al., "Assessing Effects of l-Methylfolate in Depression Management: Results of a Real-World Patient Experience Trial," *Primary Care Companion for CNS Disorders* 15, no. 4 (2013): PCC.13m01520, https://doi.org/10.4088/PCC .13m01520.

191 the FDA has approved them: Sanjay J. Mathew, "Treatment-Resistant Depression: Recent Developments and Future Directions," *Depression and Anxiety* 25, no. 12 (2008): 989–92, https://doi.org/10.1002/da.20540; "Deplin Medical Food Provides Long-Term Benefit for Major Depressive Disorder," MPR, May 8, 2012.

193 symptoms of bipolar disorder do occur in children: "Bipolar Disorder in Children and Teens," National Institute of Mental Health, revised 2020, https://www.nimh.nih.gov/health/publications/bipolar-disorder-in-children-and-teens/index.shtml; "Bipolar Disorder in Children," Harvard Mental Health Letter, Harvard Health Publishing, May 2007, https://www.health.harvard.edu/newsletter_article/Bipolar_disorder_in_children.

194 A double mutation in this gene: Damali N. Martin, et al., "Association of MTHFR Gene Polymorphisms with Breast Cancer Survival," *BMC Cancer* 6, no. 257 (October 2006), https://doi.org/10.1186/1471-2407-6-257.

194 children who experience early trauma: Nicole R. Nugent et al., "Gene-Environment Interactions: Early Life Stress and Risk for Depressive and Anxiety Disorders," *Psychopharmacology* 214, no.1 (2011): 175–96, https://doi.org/10.1007/s00213-010-2151-x; Gustavo Turecki and Michael Meaney, "Effects of the Social Environment and Stress on Glucocorticoid Receptor Gene Methylation: A Systematic Review," *Biological Psychiatry* 79, no. 2 (2016): 87–96, https://doi.org/10.1016/j.biopsych.2014.11.022.

195 prime candidate for a mood disorder: A. Lok et al., "Interaction Between the MTHFR C677T Polymorphism and Traumatic Childhood Events Predicts Depression," *Translational Psychiatry* 3, no. 7 (July 2013): e288, https://doi.org/10.1038/tp.2013.60; Torsten Klengel et al., "The Role of DNA Methylation in Stress-Related Psychiatric Disorders, *Neuropharmacology* 80 (May 2014): 115–32, https://doi.org/10.1016/j.neuropharm.2014.01.013; S. J. Lewis et al., "The Thermolabile Variant of MTHFR Is Associated with Depression in the British Women's Heart and Health Study and a Meta-Analysis," *Molecular Psychiatry* 11, no. 4 (2006): 352–60, https://doi.org/10.1038/sj.mp.4001790; Christopher B. Kelly et al., "The MTHFR C677T Polymorphism Is Associated with Depressive Episodes in Patients from Northern Ireland," *Journal of Psychopharmacology* 18, no. 4 (2004): 567–71, https://doi.org/10.1177/0269881104047285; Lin Wan et al. "Methylenetetrahydrofolate Reductase and Psychiatric Diseases," *Translational Psychiatry* 8, no. 1 (November 2018): 242, https://doi.org/10.1038/s41398-018-0276-6.

195 hindering the body's ability: Lok et al., "Interaction Between the MTHFR C677T."

195 enables the production of glutathione: Jeremy W. Gawryluk et al., "Decreased Levels of Glutathione, the Major Brain Antioxidant, in Post-Mortem Prefrontal Cortex from Patients with Psychiatric Disorders," *International Journal of Neuropsychopharmacology* 14, no. 1 (February 2011): 123–30, https://doi.org/10.1017/S1461145710000805.

195 low Vitamin D: Pei Li et al., "Vitamin D Deficiency Causes Defective Resistance to *Aspergillus fumigatus* in Mice via Aggravated and Sustained Inflammation," *PLOS One* 9, no. 6 (June 2014): e99805, https://doi.org/10.1371/journal.pone.0099805.

195 a common finding in kids with ITABI: Gonca Çelik et.al, "Vitamin D Deficiency in Obsessive-Compulsive Disorder Patients with Pediatric Autoimmune Neuropsychiatric Disorders Associated with Streptococcal Infections: A Case Control Study," *Nöropsikiyatri arşivi* 53, no. 1 (March 2016): 33–37, https://www.ncbi.nlm.nih.gov/pubmed/28360763.

195 vitamin D seems to improve symptoms: Hong-Hua Liet al., "Clinical Improvement Following Vitamin D3 Supplementation in Children with Chronic Tic Disorders," *Neuropsychiatric Disease and Treatment* 15 (August 2019): 2443–50, https://doi.org/10.2147/NDT.S212322.

195 Vitamin D also helps protect the integrity: Shiori Takahashi et al., "Active Form of Vitamin D Directly Protects the Blood-Brain Barrier in Multiple Sclerosis," *Clinical and Experimental Neuroimmunology* 8, no. 3 (August 2017): 244–54, https://doi.org/10.1111/cen3.12398; Iqbal Sayeed, Nefize Turan, Donald G Stein, Bushra Wali, "Vitamin D Deficiency Increases Blood-Brain Barrier Dysfunction After Ischemic Stroke in Male Rats," *Experimental Neurology* 312 (2019): 63–71, doi:10.1016/j.expneurol.2018.11.005; Soonmi Won et al., "Vitamin D Prevents Hypoxia/Reoxygenation-Induced Blood-Brain Barrier Disruption via Vitamin D Receptor-Mediated NF-kB Signaling Pathways," *PLoS One* 10, no. 3 (March 2015): e0122821, https://doi.org/10.1371/journal.pone.0122821; Budbazar Enkhjargal et al., "Intranasal Administration of Vitamin D Attenuates Blood-Brain Barrier Disruption Through Endogenous Upregulation of Osteopontin and Activation of CD44/P-Gp Glycosylation Signaling After Subarachnoid Hemorrhage in Rats," *Journal of Cerebral Blood Flow & Metabolism* 53, no. 1 (March 2016): 33–37, https://doi.org/10.1177/0271678X16671147.

196 give her body what it needed: "Deplin Medical Food."

196 increase glutathione: S. Jill James et al., "Efficacy of Methylcobalamin and Folinic Acid Treatment on Glutathione Redox Status in Children with Autism," *American Journal of Clinical Nutrition* 89, no. 1 (January 2009): 425–30, https://doi.org/10.3945/ajcn.2008.26615.

196 It's also been used in treatments for chronic fatigue syndrome: Health Watch, "Methylcobalamin: A Potential Breakthrough in Neurological Disease," February 1, 1999, https://www.prohealth.com/library/methylcobalamin-a-potential-breakthrough-in-neurological-disease-2-11594.

196 has been linked to depression: G. B. Parker, H. Brotchie, and R. K. Graham, "Vitamin D and Depression," *Journal of Affective Disorders* 208 (January 2017): 56–61. doi: 10.1016/j.jad.2016.08.082.

196 causes an increase in inflammatory factors: Brain Research Lab, "Progesterone in Stroke with Comorbid Vitamin D Deficiency," Emory University School of Medicine, 2018, https://med.emory.edu/departments/emergency-medicine /innovation-discovery/brain-research/stroke-d-deficiency-progesterone.html.

196 has been shown to mitigate: Negin Masoudi Alavi et al., "Effect of Vitamin D Supplementation on Depression in Elderly Patients: A Randomized Clinical Trial," *Clinical Nutrition* 38, no. 5 (September 19, 2018): 2065–70, https:// doi.org/10.1016/j.clnu.2018.09.011.

196 as well as depression in teenagers: Tara Haelle, "Vitamin D Levels Linked to Depression in Teens," *Pediatric News*, May 17, 2019, https://www.mdedge .com/pediatrics/article/201186/mental-health/vitamin-d-levels-linked-depression -teens.

196 vitamin E: S. Tidow-Kebritchi and S. Mobarhan, "Effects of Diets Containing Fish Oil and Vitamin E on Rheumatoid Arthritis," *Nutrition Review* 59, no. 10 (2001): 335–38, doi:10.1111/j.1753-4887.2001.tb06958.x.

201 low-lying depression and perfectionistic: Judy Tsafrir, "MTHFR, Methylation and Histamine in Psychiatric Conditions," *Psychology Today*, November 22, 2017, https://www.psychologytoday.com/us/blog/holistic-psychiatry/201711 /mthfr-methylation-and-histamine-in-psychiatric-conditions.

201 SAMe has been shown to augment: J. Craig Nelson, "S-Adenosyl Methionine (SAMe) Augmentation in Major Depressive Disorder," *American Journal of Psychiatry* (August 2010), https://doi.org/10.1176/appi.ajp.2010.10040627.

202 most Americans have insufficient Vitamin D: Sue Penckofer et al., "Vitamin D and Depression: Where Is All the Sunshine?" *Issues in Mental Health Nursing* 31, no. 6 (2010): 385–93, doi:10.3109/01612840903437657.

202 optimal levels of Vitamin D: Jonathan A. Shaffer et al., "Vitamin D Supplementation for Depressive Symptoms: A Systematic Review and Meta-analysis of Randomized Controlled Trials," *Psychosomatic Medicine* 76, no. 3 (2014):190–96, doi:10.1097/PSY.0000000000000044; Vikas Menon et al., "Vitamin D and Depression: A Critical Appraisal of the Evidence and Future Directions," *Indian Journal of Psychological Medicine* 42, no. 1 (January 2020):11–21, https://doi.org/10.4103/IJPSYM.IJPSYM_160_19; Deborah Brauser, "Diet and Mental Health: The Evidence to Date," Medscape, January 15, 2020, https://www.medscape.com/viewarticle/923817.

204 vitamin D receptor: Iffat Hassan et al., "Association of Vitamin D Receptor Gene Polymorphisms and Serum 25-Hydroxy Vitamin D Levels in Vitiligo: A Case-control Study," *Indian Dermatology Online Journal* 10, no. 2 (2019): 131–38, doi:10.4103/idoj.IDOJ_97_18; Maryam Mukhtar et al., "Vitamin D. Receptor Gene Polymorphism: An Important Predictor of Arthritis Development," *BioMed Research International* 2019, article ID 8326246 (March 2019), https://doi.org/10.1155/2019/8326246.

204 symptoms of overmethylation include: Judy Tsafrir, "MTHFR."

204 has been shown to restore: Georgetown University Medical Center, "Resveratrol Appears to Restore Blood-Brain Barrier Integrity in Alzheimer's Disease," ScienceDaily, July 27, 2016, www.sciencedaily.com/releases/2016/07/160727140041.htm.

205 "take home" lead: Occupational Safety and Health Administration, "If You Work Around Lead, Don't Take It Home!" OSHA.gov, https://www.osha.gov/Publications/OSHA3680.pdf.

205 In 2016, the World Health Organization: A. Prüss-Ustün et al., *Preventing Disease Through Healthy Environments: A Global Assessment of the Burden of Disease From Environmental Risks* (Geneva: World Health Organization, 2016), x. Updated figures may be accessed at https://www.who.int/quantifying_ehimpacts/publications/preventing-disease/en/.

205 Thanks to better environmental regulations: Ed Yong, "Some Fish Are Still Full of Mercury, for a Worrying Reason," *The Atlantic*, August 7, 2019, https://www.theatlantic.com/science/archive/2019/08/why-changing-climate-means-more-mercury-seafood/595663/; Yanxu Zhang et al., "Observed Decrease in Atmospheric Mercury Explained by Global Decline in Anthropogenic Emissions," *Proceedings of the National Academy of Sciences* 113, no. 3 (January 2016): 526–31, https://doi.org/10.1073/pnas.1516312113.

205 atmospheric mercury levels in the U.S.: David G. Streets, et al., "Global and Regional Trends in Mercury Emissions and Concentrations, 2010–2015," *Atmospheric Environment* 201 (March 2019): 417–27, https://doi.org/10.1016/j.atmosenv.2018.12.031.

205 30 percent decline: Yong, "Some Fish."

205 polyester, which is naturally flame resistant: Elisabeth Leamy, "How to Find Flame-Resistant Pajamas for Kids, Without Toxic Chemicals," *Washington Post*, November 16, 2017, https://www.washingtonpost.com/lifestyle/on-parenting/how-to-find-flame-resistant-pajamas-for-kids-without-toxic-chemicals/2017/11/08/fe587216-c32d-11e7-afe9-4f60b5a6c4a0_story.html.

205 Clothing manufacturers are required: Children's Sleepwear Regulations, "What Requirements Apply to My Product," CPSC.gov, https://www.cpsc .gov/Business--Manufacturing/Business-Education/Business-Guidance /Childrens-Sleepwear-Regulations.

205 the requirements for a tight fit: Elisabeth Leamy, "How to Find Flame-Resistant Pajamas."

206 And yet as rapidly developing countries: Streets et al., "Global and Regional Trends."

206 The FDA found mercury in: Walter J. Crinnion, "Environmental Medicine, Part Three: Long-Term Effects of Chronic Low-Dose Mercury Exposure," *Alternative Medicine Review* 5, no. 3 (2000): 209–23, PMID: 10869102; U.S. Food and Drug Administration, "Mercury Levels in Commercial Fish and Shellfish (1990–2012)," FDA.gov, October 2017, https://www .fda.gov/food/metals-and-your-food/mercury-levels-commercial-fish-and -shellfish-1990-2012?.

206 mercury has been linked: J. Lebel, D. Mergler, and M. Lucotte, "Evidence of Early Nervous System Dysfunction in Amazonian Populations Exposed to Low-Levels of Methylmercury," *Neurotoxicology* 17, no. 1 (Spring 1996): 157–67, PMID: 8784826; Stephen Genuis, "Toxicant Exposure and Mental Health-Individual, Social, and Public Health Considerations," *Journal of Forensic Sciences* 54 (2009): 474–77, https://doi.org/10.1111 /j.1556-4029.2008.00973.x; Xuebing Huang et al., "Mercury Poisoning: A Case of a Complex Neuropsychiatric Illness," *American Journal of Psychiatry*, 171 no. 12 (December 2014): 1253–56, https://doi.org/10.1176/appi.ajp .2013.12101266.

206 Low levels of lead and cadmium: Orish Ebere Orisakwe, "The Role of Lead and Cadmium in Psychiatry," *North American Journal Of Medical Sciences* 6, no. 8 (2014): 370–76, doi:10.4103/1947-2714.139283.

206 There's evidence that prolonged: Howard Hu, James Shine, and Robert O. Wright, "The Challenge Posed to Children's Health by Mixtures of Toxic Waste: The Tar Creek Superfund Site as a Case-Study," *Pediatric Clinics of North America* 54, no. 1 (February 2007): 155–75, https://doi.org/10.1016/j .pcl.2006.11.009; Virginia A. Rauh and Amy Margolis, "Research Review: Environmental Exposures, Neurodevelopment, and Child Mental Health: New Paradigms for the Study of Brain and Behavioral Effects," *Journal of Child Psychology and Psychiatry* 57 no. 7 (July 2016): 775–93, https://doi .org/10.1111/jcpp.12537.

206 Exposure to lead in early childhood: Rauh and Margolis, "Research Review."

207 chipped walls, lead-contaminated dust: National Center for Environmental Health, "Prevent Children's Exposure to Lead," Centers for Disease Control, CDC.gov, last reviewed October 21, 2019, https://www.cdc.gov/nceh/features /leadpoisoning/index.html.

207 Electromagnetic frequencies: Ju Hwan Kim et al., "Possible Effects of Radio-frequency Electromagnetic Field Exposure on Central Nerve System," *Biomolecules & Therapeutics* 27, no. 3 (2019): 265–75, https://doi.org/10.4062 /biomolther.2018.152.

207 Dramatic impact on metabolism: L. D. Empting, "Neurologic And Neuro-psychiatric Syndrome Features of Mold And Mycotoxin Exposure," *Toxicology and Industrial Health* 25, no. 9–10, (October–November 2009): 577–581, https://doi.org/10.1177/0748233709348393; Judy Tsafrir, "Mold Toxicity: A Common Cause of Psychiatric Symptoms," *Psychology Today*, August 3, 2017, https://www.psychologytoday.com/us/blog/holistic-psychiatry/201708 /mold-toxicity-common-cause-psychiatric-symptoms.

209 which has been known to trigger: Jennifer Frankovich et al., "Five Youth with Pediatric Acute-Onset Neuropsychiatric Syndrome of Differing Etiologies," *Journal of Child and Adolescent Psychopharmacology* 25, no. 1 (2015): 31–37, https://doi.org/10.1089/cap.2014.0056.

212 The mitochondria produce 95 percent: Gislaine T. Rezin, Graziela Amboni, et al., "Mitochondrial Dysfunction and Psychiatric Disorders," *Neurochemical Research* 34, no. 1021 (2009): https://doi.org/10.1007/s11064-008-9865-8.

213 Primary mitochondrial dysfunction is caused: D. A. Rossignol and R. E. Frye, "Mitochondrial Dysfunction in Autism Spectrum Disorders: A Systematic Review and Meta-Analysis," *Molecular Psychiatry* 17, no. 3 (2012): 290–314, https://doi.org/10.1038/mp.2010.136; Susanne Michels et al., "Downregulation of the Psychiatric Susceptibility Gene *Cacna1c* Promotes Mitochondrial Resilience to Oxidative Stress in Neuronal Cells," *Cell Death Discovery* 4, no. 54 (May 2018), https://doi.org/10.1038/s41420-018-0061-6.

213 the mechanisms that make ATP require: E. Wesselink et al., "Feeding Mitochondria: Potential Role of Nutritional Components to Improve Critical Illness Convalescence," *Clinical Nutrition* 38, no. 3 (August 2018): 982–95, https://doi.org/10.1016/j.clnu.2018.08.032.

213 prenatal infections or complications: Michels et al., "Downregulation."

213 the brain's energy requirements are the most: Lauren Owen and Sandra I. Sunram-Lea, "Metabolic Agents That Enhance ATP Can Improve Cognitive

Functioning: A Review of the Evidence for Glucose, Oxygen, Pyruvate, Creatine, and L-Carnitine," *Nutrients* 3, no. 8 (2011): 735–55, https://doi.org/10.3390/nu3080735.

213 low ATP production leads to neuronal death: Rezin, Amboni, et al., "Mitochondrial Dysfunction"; Keiko Iwata, "Mitochondrial Involvement in Mental Disorders; Energy Metabolism, Genetic, and Environmental Factors," *Methods in Molecular Biology* 1916 (2019): 41–48, https://doi.org/10.1007/978-1-4939-8994-2_2.

213 evidence is building: M. E. Breuer et al., "The Role of Mitochondrial OXPHOS Dysfunction in the Development of Neurologic Diseases," *Neurobiology of Diseases* no. 51 (2013): 27–34, https://doi.org/10.1016/j.nbd.2012.03.007.

213 A combination of genetic and environmental: Michels, et al., "Downregulation," 6; Wesselink et al., "Feeding Mitochondria."

213 mitochondrial dysfunction can affect cognition: Hayley Clay, Stephanie Sillivan, and Christine Konradi, "Mitochondrial Dysfunction and Pathology in Bipolar Disorder and Schizophrenia," *International Journal of Developmental Neuroscience: The Official Journal of the International Society for Developmental Neuroscience* 29, no. 3 (2011): 311–24. https://doi.org/10.1016/j.ijdevneu.2010.08.007.

213 dysfunctions in this process: Rezin et al., "Mitochondrial Dysfunction"; Iwata, "Mitochondrial Involvement."

213 Mitochondrial dysfunction has been implicated: Kalyan Reddy Manda et al., "Highly Active Antiretroviral Therapy Drug Combination Induces Oxidative Stress and Mitochondrial Dysfunction in Immortalized Human Blood-Brain Barrier Endothelial Cells," *Free Radical Biology and Medicine* 50, No. 7 (April 2011): 801–10, https://doi.org/10.1016/j.freeradbiomed.2010.12.029; Iwata, "Mitochondrial Involvement."

213 linked mitochondrial disorders to mental health disorders: Rebecca E. Anglin et al., "The Psychiatric Manifestations of Mitochondrial Disorders: A Case and Review of the Literature," *Journal of Clinical Psychiatry* 73, no. 4 (2012): 506–12, https://doi.org/10.4088/JCP.11r07237; Rezin, et al. "Mitochondrial Dysfunction"; Shaw-Hwa Jou, Nan-Yin Chiu, and Chin-San Liu, "Mitochondrial Dysfunction and Psychiatric Disorders," *Chang Gung Medical Journal* 32, no. 4 (2009): 370–79; Sumana Chakravarty et al., "Chronic Unpredictable Stress (CUS)-Induced Anxiety and Related Mood Disorders in a Zebrafish Model: Altered Brain Proteome Profile Implicates Mitochondrial Dysfunction," *PlosOne* (May 2013), https://doi.org/10.1371/journal.pone.0063302.

Index

About the Author

Kenneth A. Bock, M.D., is a board-certified physician who received his medical degree with honor from the University of Rochester School of Medicine in 1979. An internationally recognized pioneer of integrative medicine, he is the bestselling author of *Healing the New Childhood Epidemics, The Road to Immunity, Natural Relief for Your Child's Asthma*, and *The Germ Survival Guide*. He has also contributed to numerous professional publications and is an in-demand national and international speaker. Over the course of his thirty-five-year career, he has become known for his unique ability to identify and untangle the most complex, multisystem, multisymptom medical cases. His world-renowned private practice, Bock Integrative Medicine, is located in Red Hook, New York, in the beautiful Hudson Valley.